Novel Extension Approaches for Reshaping Indian Agriculture

The Editors

Dr. U.S. Gautam is Vice Chancellor, Banda University of Agriculture and Technology, Banda, (UP). He was awarded with Ph.D in Dairy Extension from National Dairy Research Institute, Karnal in 1989. He has experience of about 30 years in research, extension and training in the extension. Dr. Gautam began his career as a Scientist in KVK, Kullu in 1989. He has served as Principal Scientist at ICAR Research Complex for Eastern Region, Patna. He has also served ICAR-ATARI, Jabalpur and ICAR-ATARI, Kanpur as Director. He has handled several national and international multi-disciplinary projects focusing on Improvement of livelihood of the small and marginal farmers and farm women. He has published 12 books, several book chapters and more than 60 research papers in reputed national and international journals. He is a recipient of several awards including 'Fakruddin Ali Ahmed award 2010', 'Swamy Sahajanand Saraswati Extension Scientist Award' and 'ICAR Award for Outstanding Multidisciplinary Team Research in Agricultural and Allied Science'.

Dr. Manjeet Singh Nain, Principal Scientist, Agril. Extension, ICAR-IARI, New Delhi graduated from CCSHAU Hisar and obtained his Master and Doctorate degree from GB Pant university of Agri. & Technology, Pantnagar. He has 23 years of professional experience in teaching, research and technology transfer. He started his professional career during 1998 from KVK Leh (J&K), later shifted to Jammu and worked for 12 years at Sher e Kashmir University of Agricultural Sciences and Technology of Jammu in the capacity of Assistant Professor and Associate Professor (Agricultural Extension). He has published more than 160 papers including 125 NAAS rated research papers, 30 book chapters, nine books, technical papers and popular articles for farmers. He has guided 12 research students (M. Sc. and Ph.D.) as chairman and 18 as co-chairman /member. His area of research includes; agricultural communication, diffusion of innovation, participatory methods, entrepreneurship development farmer led innovations and training methodologies. He has handled 17 research projects including 10 outside funded projects. He is recipient of Young Scientist Award of Society of Extension Education, Mobilization Fellow, Best Community Mobilizer, Fellow of Indian Society of Extension Education, Dr. K.N. Singh memorial award of Indian Society of Extension Education, Ram Nath Singh Award of ICAR-IARI, New Delhi, in addition to other appreciations. He is Chief Editor of Indian Journal of Extension Education, and Journal of Extension System and member of editorial board of Journal of Community Mobilization and Sustainable Development.

Dr. Dheeraj Mishra did his Ph.D. in Extension Education from Banaras Hindu University, Varanasi. He has more than 5 years of experience in teaching and research. He has published more than 23 research papers/book chapters and popular articles in Journals and Magazines of National and International repute. He is a life member of various academic societies. Presently Dr. Mishra is working as Assistant Professor in Department of Agricultural Extension, Banda University of Agriculture and Technology, Banda.

Dr. Vister Joshi is presently working as Subject Matter Specialist (Agriculture Extension) in Krishi Vigyan Kendra, Jalaun under the jurisdiction of Banda University of Agriculture and Technology, Banda, Uttar Pradesh. With wide experience in rendering extension services, Dr. Joshi also work to assess the Impact of agricultural technologies during his tenure at ICAR-National Institute of Agriculture Economics and Policy Research, New Delhi. He also served the reputed The Energy and Resources Institute (TERI), New Delhi under the integrated pest management project in collaboration with IPL-Innovative lab, Virginia Tech University, USA. He also had been in services of eminent Agriculture Today Group at New Delhi. Dr. Joshi did his Ph.d in Extension Education from CCSHAU, Hisar, and Masters from PAU, Ludhiana and Graduated from GBPUAT, Pantnagar. He has been granted with ICAR- NTES, ICAR-JRF during academics. Dr. Joshi has published many research papers, books, and articles in different professional Daises.

Novel Extension Approaches for Reshaping Indian Agriculture

– Editors –

Dr. U.S. Gautam

Vice Chancellor,
Banda University of Agriculture and Technology, Banda

Dr. M.S. Nain

Principal Scientist
Division of Agricultural Extension
ICAR-Indian Agricultural Research Institute, Pusa, New Delhi

Dr. Dheeraj Mishra

Assistant Professor, Agricultural Extension,
Banda University of Agriculture and Technology, Banda

Dr. Vister Joshi

Subject Matter Specialist,
Agriculture Extension, Krishi Vigyan Kendra, Jalaun

BIOTECH

2022

BIOTECH BOOKS®

ISBN 978-81-7622-514-4

This book is a compilation of lead presentations delivered during ISEE National Seminar 2019 held at CCSHAU, Hisar during 20-22 November, 2019 organised by Indian Society of Extension Education, IARI, New Delhi.

Published by: **BIOTECH BOOKS**®
4762-63/23, Ansari Road, Darya Ganj,
New Delhi - 110 002
Phone: 09899295259, 011-43003222
E-mail: biotechbooks@yahoo.co.in
Website: www.biotechbooks.in

Typeset at: **Classic Computer Services**
Delhi - 110 035

Printed at: **Replika Press Pvt. Ltd.**

PRINTED IN INDIA

रानी लक्ष्मी बाई केन्द्रीय कृषि विश्वविद्याल
ग्वालियर रोड, झांसी 284 003 (उत्तर प्रदेश) भारत
Rani Lakshmi Bai Central Agricultural Universit
Gwalior Road, Jhansi 284 003 (U.P.) India

प्रोफ़ेसर पंजाब सिंह
कुलाधिपति
Prof. Panjab Singh
Chancellor

Phone : 0542-2575678 (R)
Mobile : 9651666613
E-mail : panjabsingh03@yahoo.co
Website : www.rlbcau.ac.in

Foreword

The agriculture sector is critical for livelihoods and employment with the engagement of 54.6 percent population of the country. Along with the agricultural production, financial prosperity of the farmers is also a major challenge that needs innovations, technologies and good management practices. The extension system has come up with new approaches with multiple forms, actors, and structures. These approaches involve digitally-enabled tools or services for bridging the existing gaps. The recent technological advancements in GIS, remote sensing, experts systems, decision support systems, e-knowledge gates and e-marketing portals created a number of digital platforms to provide various regulatory and technical solutions to agricultural production.

Farming sideways driven by market, create equally difficult challenge for weak extension system which is surrounded by the more innovative competitors in the information marketplace. Moreover, new social initiatives of Farmer Producer organizations, market federations, and farmer interest groups intended to fetch market access with better prices. Extension management to fruit these social experiments without digital platform comes under consideration amongst the policy frame workers. The socio-digital blend could ensure paradigm shifts in technology generation, their dissemination and adoptions ensures social inclusion and also open limitless possibilities for researches in agricultural sciences.

The book has presented an overview of recent socio-digital approaches for sustainable agricultural developments. It contains chapters on community based organizations to empower rural community, innovative extension research and policy, entrepreneurship development and technological interventions for climate smart agriculture. These contributions are made by the eminent extension scientists.

Former Secretary DARE, Govt. of India and DG, ICAR, New Delhi & Vice-Chancellor, Banaras Hindu University, Varana:

Significant contributions have emphasized on food and nutritional security as well as gender issues in agriculture. This could be possible when there is understanding of socio-economic values of agro-ecosystem services and a mechanism to converge it on digital ground. Further actions on these lines shall take the concept forward and incentivize the extension fraternity to render their services for farmers' well-being. It is hoped that this book shall encourage advancement in digital agriculture extension.

I congratulate the Indian Society of Extension Education, IARI, New Delhi for bringing out this edited book on '*Novel Extension Approaches for Reshaping Indian Agriculture*'. The tireless efforts of Dr. Gautam and other editors of the book are much appreciated for this academic contribution.

(Panjab Singh)

Preface

COVID-19 pandemic has caused significant human suffering and extensively disrupted the Indian agricultural system. The pandemic wreaked a substantial physical, social, economic and emotional havoc on all the stakeholders of Indian agricultural system. Sustenance of an optimum level of agricultural productivity requires efficient and timely use of inputs, safety and comfort of the agricultural worker, improvement in the quality and value addition to the produce. Also, profoundly affected production and marketing through labor and logistical constraints, while the negative income shock restricted access to markets and increased prices of food commodities affecting the consumption pattern which has long lasting socio-economic consequences.

As every crisis is an opportunity by itself. Extension provides a critical support service for rural producers meeting the new challenges confronting agriculture. Since, Agricultural extension plays a crucial role in boosting agricultural productivity, increasing food security, improving rural livelihoods, and promoting agriculture as an engine of pro-poor economic growth. With the changing scenario a newer range of extension approaches which include farmer to farmer extension, innovative way of solving farmers' need using ICTs and extension network, market-led extension, entrepreneurship came out as core aspects to reinvigorate the agriculture in the nation. Evolutionary extension can greatly make use of ICTs effectively to respond to the information and other needs of farmers.

The present attempt is a treasury to extension fraternity as the newer insight in the system set down to develop the socio-economic situation of the farming community. The métier of this edition is the secrets from the stalwarts of the Indian Extension System, which is based on the experiences and newer finding in the discourse during the national seminar. The finding covers almost all dimensions of extension to up bring the livelihood of farmers through better advisories through remarkable system. During the course of this work, the information provided by researchers and experts edited for brevity and relevance. It is firm believe that this compilation would be instrumental for the researchers, extension personnel's, policy makers, students and all farmer well-wishers resulting in the increased awareness about the innovative accrescent extension system.

Editors

Contents

Chapter 1

Innovative Agricultural Extension Reforms for Efficient Programme Delivery: Challenges and Way Ahead

V.V Sadamate[1], R.N Padaria[2] and Priti Priyadarshni[3]

[1]Former Adviser (Agriculture), then Planning Commission, GOI, New Delhi
[2]Professor and [3]Ph.D.Scholar, Division of Agricultural Extension, IARI, New Delhi – 110012

1. Agriculture and Allied Sectors: Impressive Performance

1. The performance of farm sector in India is quite impressive (Annual Report DAC and FW, 2017-18) in terms of its achievements in crops and allied sectors over the years. Food production has touched a new height of 284.83 Million Tonnes (MTs), with record score of pulses (25.23 MTs) and oil seeds (31.31MTs) production.
2. The fruit (100 MTs) and vegetable (166MTs) sectors have recorded significant achievements (DAC and FW, 2018).
3. India is now the World's largest producer of milk (176 MTs) with per capita availability of 337 grams as against the Word average of 300 grams (Annual Report, DAHD and F, 2018-19).
4. The fish production has touched the level of 11 MTs with 7 MTs of inland and 4 MTs of marine (Annual Report, DAHD and F, 2018-19).
5. The meat production in the country is at 10.5 MTs whereas the egg production has gained a record achievement of 90 Billion eggs per annum with per capita consumption of 63 eggs per annum (Annual Report, DAHD and F, 2018-19).

6. Like other crucial inputs, the technology dissemination services, both public and private have played a significant role in the achievements cited above. Though the performance of extension delivery varied from State to State.

2. Farm Sector: Current Concerns

Agriculture Sector as a whole faces formidable challenges, despite record achievements. These include: shrinking land and water resources, adverse impact of climate change, skewed development in various agro-climatic zones, emergence of farm distress though sporadic, stagnant productivity in some crops/areas, increased production costs, dwindling natural resources, market uncertainties and need for reforms in technology delivery systems across States and sectors. The major concerns revolve around the following.

2.1 GDP Contribution of Farm Sector Sliding Down

Contribution of agriculture and allied sector in GDP is sliding down (14.39 per cent at constant prices and 15.87 at current prices, MOS and PI, 2018-19); still around 60 per cent population depends on it. Allied Sectors are contributing significantly, hence need adequate programmatic and investment focus.

2.2 Shrinking Land Holdings and Serious Strain on Natural Resource Base

Predominance of small holders in Indian agriculture continues to rise. Average size of land in India has reduced to 1.30 hectare per farm holding. Small and marginal farmers now comprise of 83 per cent of total land holdings and they occupy about 41 per cent of area under all operational holdings (Agril. Statistics, 2017-18). Natural Resource base that sustains farming is under increasingly serious strain. Land, water, soil health and biodiversity resources are shrinking and deteriorating. Due to overexploitation, the ground water levels are going down even in irrigated areas in several parts of the country. Continuing imbalance in use of fertilizers and over doses of PP Chemicals has adversely affected soil health. Thus, long term sustainability demands that land and water are used judiciously and efficiently.

2.3 Regional Imbalance

Present level of productivity of most of the crops and other sectors is significantly low as compared to what is attainable and also are being attained by some other countries. This production potential is believed to be very large, for example, in the eastern and central regions. Regional imbalances in growth and productivity are high in rainfed conditions. Dryland areas that are home to a large section of marginalized communities continue to be in disadvantageous position in terms of desired development interventions.

2.4 Sustaining Food and Nutritional Security

India ranks at 103rd as per Global Hunger Index (IFPRI, 2018) amongst 119 nations, Nepal (72) and Bangladesh (86) are ahead of us. Half of the children (51 per cent) in India are malnourished, over 55 per cent married women suffer from anaemia and 30 per cent adults suffer from protein - energy malnutrition.

Food availability to BPL families and supply and distribution apparatus in disadvantaged areas along with food quality standards are basic concerns. Access to food and affordability of the impoverished families to harness the social entitlement are also issues that need greater attention to expand the income and livelihood options for such category of people. Horticulture, milk, meat, poultry and nutri-cereals may provide an excellent opportunity on nutritional improvement; however, the awareness and reach are still limited. Concerted extension efforts are needed to bring in supply chain management, ensure availability, access and affordability (AAA).

Malnutrition particularly with reference to essential minerals is a major problem, which can be addressed by inclusion of coarse cereals and millets in PDS, food fortification and propagation of nutrient-dense foods. Farm women could play an important role in managing and sustaining house hold food security and extension services need to be oriented that way.

2.5 Rainfed Farming and its Sustainability in the Context of Climate Change

Nearly 60 per cent cultivable area would continue to be rain fed - typically characterized by vulnerable natural resource base, inhabiting 40 per cent population, giving 40 per cent food grain output and sustaining about 60 per cent of livestock. Watershed has been taken up as a key intervention for development of such areas. Climate change and adaptation are dominant issues in farm sustainability (AESA,2019). Better convergence and programme delivery below of the sub-district levels are urgently called for. Recent innovative programmes like NICRA, NMSA and PMKSY have demonstrated good potential that need to be main-streamed into the future strategies. Technology transfer models adopted so far have not been able to make any perceptible dent in rain fed areas. It calls for alternative thinking in policy framework and delivery mechanism.

2.6 The Agricultural Distress, though Sporadic

Agricultural distress reflected in some parts of the country needs to be addressed by alternate strategies as the normal programmatic formulation seems did not work there. The focus would need to be on the conservation technologies, alternate crops and cropping patterns, crop diversification, convergence and enterprise combination suiting to local resources and alternate livelihood options. It needs well defined extension strategies with focus on training of farmers and farm women for alternate livelihoods.

2.7 Efficacy of Programme Delivery and Extension Systems

Efficacy of programme delivery mechanism is another serious concern across the sectors. Both, the public and private sector extension service providers are constrained on the issues like funding, manpower, linkages (backward/forward), HRD/ICT strategies, outreach to small producers, desired impacts on the targeted beneficiaries, *etc.* Extension outreach observed to be either very weak or practically absent in sub sectors like Animal Husbandry and Dairy (AH and D) and fisheries, though these sectors are contributing significantly.

3. Growth of Agricultural Extension Programmes/Services

1. Extension services, like other crucial inputs have played a very major role during pre- and post-Green Revolution stages. The technological backstopping for this credible gain has primarily came from the NARS System. There is a huge network of the extension functionaries accounting to public sector (1.19 lakhs) and private sectors (over 1lakh) at various levels. The State Agricultural University System and the training framework in GOI and the States have played significant role in training and capacity building of extension manpower and farmers.
2. Agricultural Advisory Services, in Pre-Independent India, were operated mostly through voluntary efforts in various parts of the country. Massive agricultural development efforts, in earlier parts of post-independence period were organized through Grow More Food Campaign (GMFC) in 1947, followed by Community Development Programme (1952) and National Extension Service (NES) blocks in 1953.Subsequently,Pre-Green Revolution interventions came in, like Intensive Agricultural District Programme (IADP,1961) and Intensive Agricultural Area Programme (IAAP, 1964). Then, came the period of Green Revolution of 1967, supported by intensive extension efforts like National Demonstrations (1965), Farmers Training Centres (FTC, 1966), Small and Marginal Farmers Development Agencies (SFDA, 1971), Krishi Vigyan Kendra (KVK,1974) and Lab to Land Programme (1979).
3. Presently, the technological backstopping at the district level has been strengthened by augmenting and expanding KVK Programme in the country (714 by now) for supporting front line extension services. These are operated by the specialized teams of SMSs based at the district level promoting technology options in micro agro climatic situations covering the range of production systems. The KVK Team comprises of a Programme Coordinator and SMS(6) in various subject matters like agronomy, plant protection, extension, home science, horticulture, animal husbandry, *etc.* Each KVK has an experimental/demo farm and live production units. The major activities include: On farm testing and validation of technologies, carrying out Front Line Demonstrations, Training of farmers, farm women, farm youth and field functionaries. KVKs in turn are backed up by the SAU/ICAR network and by the Agricultural Technology Application Research Institutes (11)
4. Around the mid-seventies (1974-75), the very major Extension intervention through Training and Visit (T and V) System of Extension was implemented in Madhya Pradesh and Rajasthan on pilot basis for strengthening field extension services. This World Bank funded initiative prescribed periodic training of field functionaries, their scheduled visits to

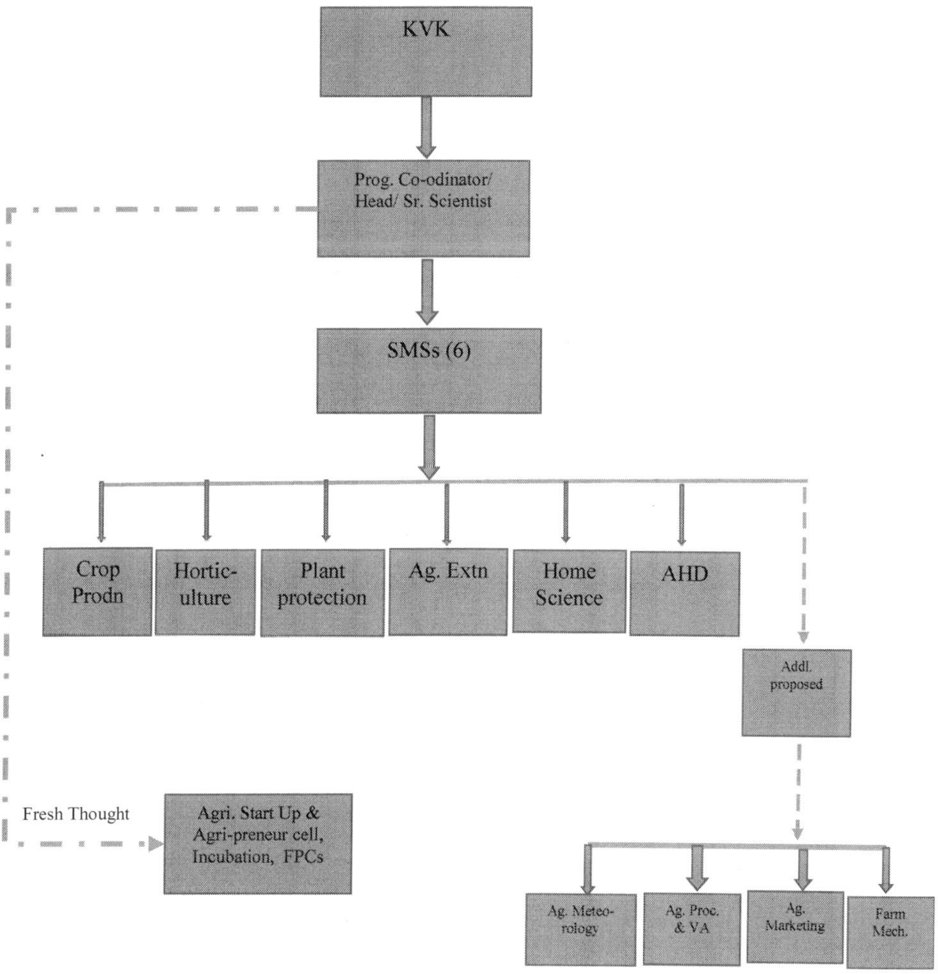

Figure 1.1: Organisational Structure of KVK (Indicating additional SMSs).

the farmers, research back up, extension training/information, mobility/infrastructure support, inbuilt M and E arrangements, *etc.* Following significant impacts, the same was expanded throughout the country from 1984 to 1995 through the World Bank supported National Agricultural Extension Project (NAEP), whereas the research back up was supported by National Agricultural Research Project (NARP).

5. Subsequently, the process of Extension Reforms was initiated under National Agricultural Technology Project (NATP, 1998) in a pilot mode with integrated extension delivery attempted under Agricultural Technology Management Agency (ATMA) at district level, now covers 676 districts in 29States and 3 UTs under National Mission on Agricultural Extension and Technology(NMAET).The Chairperson of ATMA Society is the District Collector. However, there is a full time Project Director

(PD) to execute the extension programmes and activities assisted by the Deputy PDs. At the block level there is a Block Technology Manager and three Assistant Block Technology Managers. Their field programmes are facilited through Village Extension Workers and selected Farmer Friends (1 for 2 villages). The ATMAs are supported by the block level and district level farmer committees who are involved in programme management and in providing feedback.

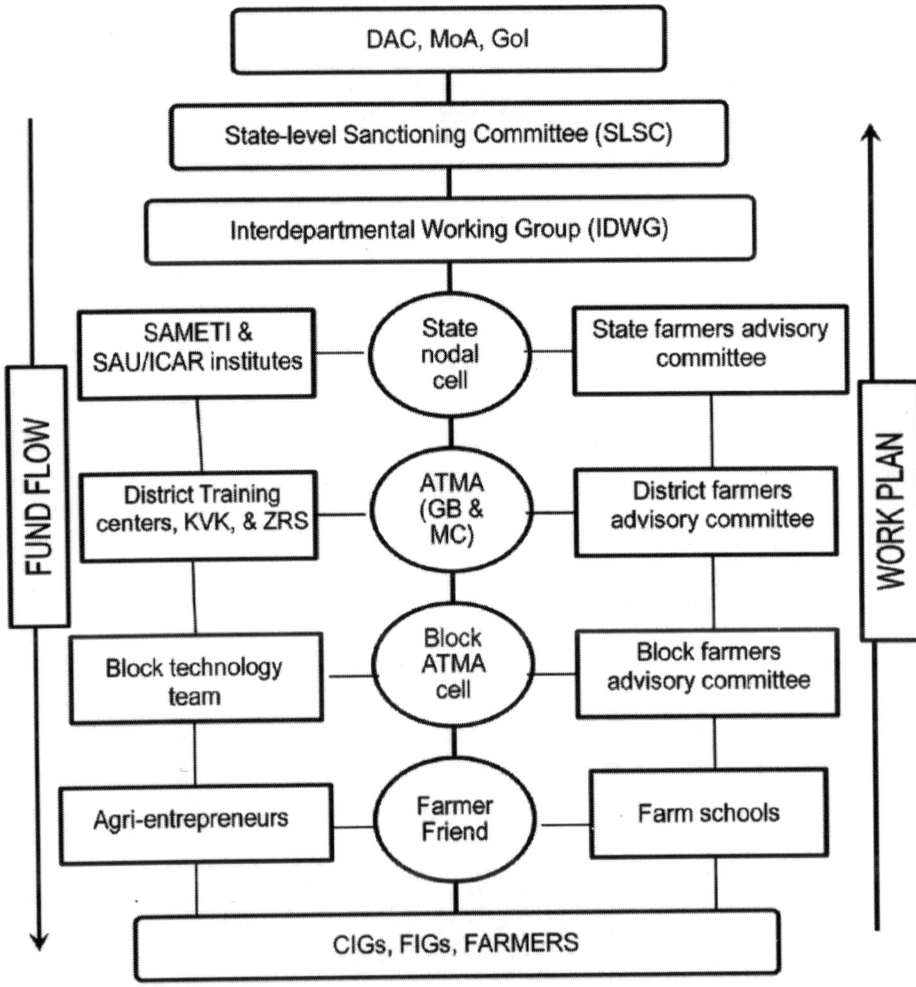

Figure 1.2: Framework of ATMA.

The field functionaries at the senior level are trained at the National Institute of Agricultural Extension Management (MANAGE) and the middle ones at State Agricultural Management and Extension Training Institutes (SAMETIs). The extension and communication trainings for field functionaries are conducted at

Regional Extension Education Institutes (EEIs) and farmers are trained at KVKs and FTCs.

Other important programmes and schemes of the Extension Division/ Directorate of Extension of the Ministry of Agriculture include: Mass Media Support to Agricultural Extension, Promoting Agri Clinic and Agri Business Centres (ACABCs), Kisan Call Centres (KCCs), National Gender Resource Centre in Agriculture (NGRCA) and National, State and Regional Exhibitions.

Major concerns: convergence platform with requisite operational flexibility, capacity to address emerging field level challenges, broad based extension delivery including crops and allied sectors.

The cafeteria of activities covered by ATMAs include: multi agency extension and convergence strategies, promoting IFS approach, gender mainstreaming and farmer centric activities like training, exposure visits, demonstrations, farm schools, organization of farmer groups, *etc.*

Emerging Pluralism in Extension Services

Agricultural extension is no longer only a public sector phenomenon. It now involves a more complex range of actors providing a wide range of services, together bracketed as extension advisory services (EAS). These include organizations in the private sector dealing with agricultural inputs, agribusiness, and financial services, non-governmental organizations (NGOs international as well as local); producer groups, cooperatives and associations; consultants (independent as well as associated with or employed by agri-business/producer associations) and information and communication technology (ICT)-based services. The job expectations from extension professionals have thus changed now and now demanding various competencies.

Currently, the agricultural extension services are moving on a pluralistic platform, enabling provision of defined role space for all players. Doubling Farm Income Committee (DFI Committee, 2018) has proposed to transform agricultural extension as an important engine for doubling the farmers' income while ensuring national food and nutritional security through sustainable means.

4. Extension Models and Approaches World Over and in India

4.1 Agricultural Extension Models World over

- ✰ Knowledge Generating Systems (KGS) Specialists directly work with the farmers: West Europe, US
- ✰ Transmission from KGS - Field functionaries (SDAs) – Farmers: South Asian countries like India, Pakistan, Bangladesh, *etc.*
- ✰ Farmer Producer Organizations playing significant role in extension services: Latin America
- ✰ Farmers Associations playing significant role in Extension: South East Asia

- ☆ Transmission of Extension Messages: MOA - Field functionaries the Provinces – Farmers: Developing African countries
- ☆ Other Models. All over, Donor Driven, NGO Driven, ICT Driven (internet Platforms and Social media)

4.2 Extension Models in India

4.2.1 Public Sector Extension Service Providers (ESPs)

A. *Research Institution Outlets*

SAUs, SHUs, SVetUs ICAR/CGIAR institutes, KVKs

Primary Functions: Generation and testing of Technologies, Frontline extension services

B. *Development Department Outlets (State Sect, CS and CSS Schemes)*

Agriculture, Horticulture, Animal Husbandry Dairy and Fisheries, Sericulture *etc.*

Primary Functions: Adoption of tested technologies and Up-Scaling through schematic support.

4.2.2 Private Sector Extension Service Providers

Agri-Entrepreneurs, ACABCs, Agri Start-ups, Internet Platforms

Input Agencies, Companies and Corporates

Primary Functions: Supplement extension efforts while promoting their respective products and services.

4.2.3 Farmers/Farmer Organizations

Progressive/Innovative/Entrepreneur Farmers

Farmer Organizations, CIGs, FPOs, FPCs, *etc.*

Farmer Cooperatives (PACs/Commodity Cops)

Primary Functions: Organized farmer involvement, extend benefits to the member farmers and others

4.2.4 Non-Government Organizations (NGOS)

Directly working with the farmers

International NGOs/Mother NGOs/Field NGs

Primary Functions: Mobilization of Communities, Organizing them as SHGs, CBOs, *etc.*

4.2.5 Media/ICTs

- ☆ Print Media/Electronic media Radio/Television
- ☆ Internet based Applications and Apps
- ☆ Social Media

Primary Functions: High speed, good for creating mass awareness and for information/publicity/networks

4.3 Learning and Lessons Emerged

- ☆ Realizing that extension is process equilibrium and not mere TOT
- ☆ Realizing Pluralistic approach/Chains of ESPs
- ☆ Each has a defined role space/No substitute
- ☆ Making ESPs Supplement and Complement
- ☆ Focus on allied sectors, inadequate
- ☆ Marketing, business and income opportunities need to be brought in
- ☆ Bottom up/participatory processes in planning and implementation
- ☆ Making it respond to the emerging extension scenario at the micro agro situations
- ☆ Feedback documentation/analysis
- ☆ Operational flexibilities for convergence
- ☆ Right Extension priorities for dynamic actions
- ☆ Orienting AESs to the extension needs of farm youth and farm women
- ☆ In build M and E and correctives

5. Innovative Reforms and Strategies in Agril. Extension

5.1 Farmer Empowerment and their Intensive Use in Extension Network

5.1.1 Empowerment of Farmer Organizations (FOs)

Farmers could be empowered by various methods like organizing them into SHGs, FOs, FPCs, Coops, *etc.*, by providing training and capacity building, by making them partners in decision making and by providing role space in extension operations. Finance Minister's budget speech 2019 focused on promoting 10,000 FOs with present strength being 5000 (NARBARD 2019-20). There is need for re-orienting extension services to address extension needs of farm women and farm youths and their organized groups.

5.1.2 Pressure on Demand and Supply Chains

The organized farmer groups in the form of Farmer Organizations (FOs), Commodity Groups (CGs) and Farmer Producer Companies (FPCs) have great scope to put the adequate pressure on the demand and supply chains. Murray (2008) observed that producer organisations can be a potential solution for enhancing value addition in agricultural commodities coupled with utilizing the perishable produce.

5.1.3 Farm Youths and their Organized Groups for High Tech Farming (FY-HTF)

It has been observed that the Farm Youths and their Organized Groups are taking interest in secondary agriculture (processing, value addition, mechanization, aggregator, producer company operator, *etc.*) as opposed to primary agriculture. Extension services may need to address this dimension too as it gives farm youth the sense of social equilibrium at par with his/her urban peers.

5.1.4 Farmer Producer Companies (FPCs)

Successful and mature FOs could be converted into a Farmer Producer Companies with focus on organized approach to the extension operations and integrating their efforts with technologies and markets. FPCs have brought in income and business orientation in extension and production operations. Over 3000 FPCs are in operation in the Country and the number is growing.

5.1.5 Farmer Cooperatives, Collectives and Aggregates

Farmer Cooperatives are found to be a great success in the States like Maharashtra, Gujarat in fruits, commercial crops, vegetables and milk sectors. AMUL, MAHAGRAPES, *etc.*, are examples of merit and performance. The farmer collectives and aggregates have a major role in linking the farmers to the markets resulting in enhanced incomes.

5.1.6 Intensive Use of Farmers Field Schools (FFSs): Farmer to Farmer Extension

This is one of the effective methods of technology dissemination wherein the innovative and progressive farmers could share the technological information with the fellow farmers. FFS is an effective extension tool, practiced widely, needs to be up-scaled as it provided an opportunity for farmer to farmer extension. The Farmers Field Schools are organized on the fields of innovative farmers where they become the host and the teachers for the technology in which they have already excelled or have had a unique experience and achievements. The host farmer needs to be properly oriented about his roles and responsibilities in promoting the technologies demonstrated on his field. The fellow farmers would have to have a faith and access to the host farmer, his competence and in the results obtained on the farmers' field schools.

Priyadarshni *et al.* (2019) found utility of FFSs in enhancing social learning that included improved participation, engagement and collaboration among the farmers. It enabled development of favourable attitude towards local knowledge and building of mutual trust among the members. They stated further that when the farmers, scientists and extension agents are involved in social learning process a better understanding of risks and solutions could be created.

FFS differs from demonstration since it lays more emphasis on experimentation, science based learning, sharing of knowledge and skills in adoption of technologies. Now, the issue is how to promote, encourage and fund FFSs under various programmes. One of the methods tried by the ATMAs and KVKs was to involve such farmer innovators in the validation process by the research system and using them as training resources and disseminators.

5.1.7 Promoting Farmer Facilitators and Para Techs

There is a strong need for promoting informal sources of technology delivery like farmer friends, model farmers, lead farmers, *Kisan Sahayyaks* (farmer facilitators), *etc.* across the states/sectors, to enable faster technology transfer at the cutting edge levels. There are different models promoting farmer facilitators at the community levels. NGOs like Ramakrishna Mission have trained young boys and girls from the rural areas, built their capacities and used them as local farm volunteers. Every three months these young volunteers are brought back to the local KVKs for training and retraining. The Department of Agriculture, Andhra Pradesh have selected local practicing farmers as Model farmers (Adarsh Raytu) for technology transfer. *Kisan Sahayyaks* (Farmer facilitators) are used in the States of Uttar Pradesh, Bihar *etc.* In fisheries and in the animal husbandry sectors the States like West Bengal are promoting *Matsya Mitras*(Fish friends) and *Prani Bandhus*(Animal brothers).

These facilitators would need to be properly selected and trained for the specific task. They should be innovative, practicing farmers and should have acceptability of the community. So the selection process would need to be far fairer. He/She needs to be oriented to handle community level extension advisories and therefore, need periodic updating of their knowledge and skills. The field extension agencies like ATMAs, KVKs, NGOs, farmer organizations would need to promote the Farm Facilitators and use them at community levels.

ATMA has a formal provision to promote Farmer Friend (FF), one for two villages. He/she gets and honorarium and incentives under ATMA Scheme. The challenge is how best these resources can be groomed for technology dissemination and their capacities built and their involvement in the frontline and field extension programmes activated.

The States may consider taking a few pilots in selected blocks and systematically promote farmer facilitators (FFs) and demonstrate how effectively they could take on extension and technology transfer functions at the community level. This would largely fill up the extension gaps at the community level especially in the remote areas, hill and mountain eco-systems and tribal areas where the extension services remained weak.

5.1.8 Targeting FPOs/FPCs

It is proposed to target a FO per village and a FPC per 10 to 15 villages to be promoted jointly by the public and private extension service providers.

5.2 Enhancing Outreach of Ongoing Models (Front Line and Field Extension)

5.2.1 Enhancing Outreach of Extension Models in Operation and their interplay

There is need to improve performance of extension models in operation to their optimum levels with focus on:(i) R&D Linkages, (ii) research backstopping,(iii) widening the sectoral and area coverage, (iv) resource sharing, (v) training coordination, (vi) convergence and programme delivery, and (vii) improving penetration to the small and marginal producers.

5.2.2 Promote Chains of Service Providers

Field formations, by themselves are not able to penetrate down the line, especially to the small and the marginal farmers, and across the sectors and areas. This has resulted due mainly to inadequate numbers, their quality and limited outreach. Therefore, it would be better if extension services promote chains of extension agents, formal/informal for enhanced coverage (Figure 1.3). Agents in Extension chains may include Progressive farmers, Entrepreneurs, the FOs, FPCs, Cooperatives, start-ups, IT platforms, *etc.*

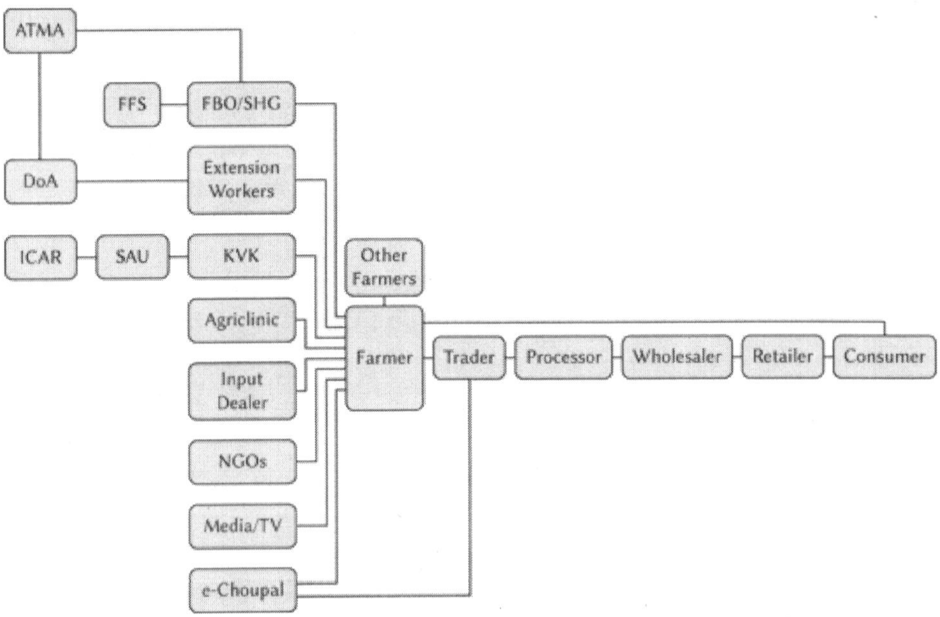

Figure 1.3: Farm Information Sources for the Farmers and Chain of Service Providers (Glendenning *et al.*, 2010).

5.2.3 Field Formations to Supplement and Complement each other

Field formations of various extension streams would need to supplement and complement in selecting programme sites, identification of beneficiaries, joint diagnosis and problem solving, information support, training and capacity building of FFs, BTMs, and ABTMs. Sharing of training resources, providing mentoring support by the KVK/SAU Scientists to the block level formations (*e.g.*, RSKs of Karnataka and *Krishi Bhavans* of Kerala) in the specialized areas may also be thought of. As above is far more applicable in case of KVK and ATMA field operations.

5.2.4 Focus, Clarity and Interplay of Front and Field Extension Management Interventions

Extension operations at the field levels need lots of clarity in appreciating finer elements of frontline (Research driven) and field extension (Development department driven). What has been demonstrated by the front line efforts at the micro environment would need to be up scaled by the field extension agencies. At times, these equations do not get properly focussed and translated into action leading to duplication of extension efforts.

5.2.5 Extension Services in Allied Sectors

With all its adequacies and inadequacies there is Agricultural Extension System exists for Crops sector. However, in allied areas like horticulture, animal husbandry, dairy, poultry, fisheries *etc.*, its either weak or missing, though the allied sectors are contributing significantly in agricultural growth. Hence, Extension needs to be re-organized thoroughly by way of (i) additional SMS support to KVKs/ATMAs in potential areas, (ii) promoting growers, breeders, fishers' groups and associations, (iii) use of milk/fish cooperatives, (iv) use of entrepreneurs, progressive farmers, *etc.*

Horticultural Extension: Growers Associations, Commodity groups, Entrepreneurs as extension agents.

AHD Extension: Besides Cattle and buffaloes, bringing in extension focus on; Sheep and Goats, Small ruminants like pigs and rabbits, CB of small producers, including farm women. Dairy Cooperatives and Commercial dairy owners as extension agents

Fisheries Extension: Focus on cooperatives, fish entrepreneurs, market integration, *etc.*

5.2.6 Market Led Extension Strategies for Enhanced Incomes

Future Extension Strategies must focus on organizing producer aggregates at various levels. This would provide strong backward and forward linkages including market integration strategies for enhanced income opportunities to the farmers. Such Farmer Aggregates would also provide adequate pressure on the R&D agencies making extension operations demand driven and participatory. Market-led extension system establishes its position by helping the farmers realize high returns for the produce, minimize the production costs and improve the product value and marketability (Kumar *et. al*, 2012).

Demand Supply projections for farm produce like grains, pulses, oilseeds, fruits, vegetables, milk and milk products, meat, fish *etc.* if worked out in advance for micro/macro situations may streamline the market behaviour and farmer may not face glut and distress sale situations.

Aggregated approach may provide proactive involvement of farmers and entrepreneurs and may provide opportunities like better price negotiating power to the farmers, pressure on provision of quality inputs and after sale services.

To realize the benefit of higher prices, farmers need to access a wider range of information, related not only to production technologies but also to postharvest processes, access to remunerative markets, price information and business development (Sulaiman and van den Ban 2003).

5.2.7 Extension for Disadvantaged Areas/Groups

Farmer distress and challenged areas need clear extension strategies/advisories including climate adaptation (Sulaiman *et al.*, 2019). It would need micro level technology options along with credit - market linkages. Appropriate enterprise combinations would need to be worked out for the blocks, clusters and the micro level agro situations to be promoted by the extension operations. This would need re-orientation of the extension machinery and improved programme delivery. Similarly, the tribal areas, the hill and mountain areas, the rainfed areas, the desert areas and the border areas would perhaps need regionally differentiated and location specific extension strategies.

Innovative extension solutions for challenged/disadvantaged areas and groups may include: (i) ESP Combinations, (ii) involving NGOs, (iii) intensive use of ICTs (iv) intensive use of Para techs, (v) participation of local communities and institutions, *etc.*

5.2.8 Benchmarks for Extension Performance and Extension Accountability

Performance standards need to be set for Extension Performance, in a given cluster, block and district. It would make the extension services accountable to the local area and farming communities, enabling extension services to respond to the emerging extension challenges: Performance benchmarks like extent of adoption, extent of convergence, enhancement of production and income of the farmers and farm women, enterprise combination, market linkages, *etc.* would need to be set for agriculture and allied sectors too.

Further accountability parameters may also include: Enhanced Penetration of extension efforts to small and marginal producers, farmer motivation on accessing extension services and putting user's pressure, attitudinal/behavioural changes obtained in a given micro ecology, farm women and youths organized and their capacities built, *etc.*

5.3 Integrating Private Sector Efforts in Extension Management

5.3.1 Bring in PPPs in Extension Management Systematically

PPPs are required down the line to promote participatory extension arrangements between the public and private extension service providers. It is expected that large number of MoU based PPPs must come up at various levels to make the extension services respond to the local situations. However, PPP operations need policy directions and clear instructions on partnerships (otherwise provisions do not get translated as it has been seen in some ATMAs). ATMAs can be a good platform for working out operational flexibilities and resource sharing, keeping in view large number of government, private and NGO extension players in the field. The input support agencies, the companies and corporate, the NGOs, the Agri Clinics and Agri Business Centres (ACABCs), *etc.*, can also play an important role in promoting and operating PPPs in Extension management covering agriculture and allied production systems (Figure 1.4).

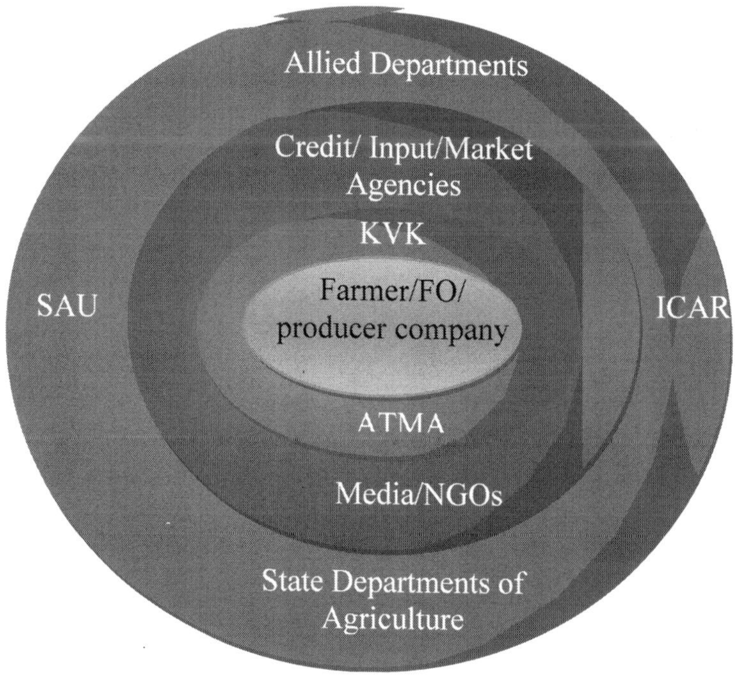

Figure 1.4: Aligning Public and Private Extension Services for Farmer Centric Approach.

A serious focus is required to understand the extension spirit of public and private extension efforts. Proper understanding of each other's efforts and mutual benefits that would be accrued is essential element in such efforts. Sometimes there is a provision; however, in the absence of field understanding such partnerships do not operate. Perhaps there is need to spell out the MOU based partnership parameters at the local level. Farmer Groups, input industries and other private extension service providers would need to be encouraged by the public extension platforms for mutual and up scaled benefits to both. Private ESPs may establish Demo units in the jurisdiction of all KVKs/ATMAs on PPP mode, promoting large number of MOU based extension operations.

PPPs in Extension involving the input dealers and dairy/fish farmers may focus on breeding, feeding and health care areas. Public sector extension agencies would need to be thoroughly oriented on importance and contribution of PPPs in promoting IFS. Special PPP Extension Provisions would need to be made under flagship programmes of the Federal and State Governments.

PPP is a pragmatic and programmatic approach to the delivery of extension services. For example, development of the value chains will require technical expertise that goes beyond the capacity of the current extension functionaries. Hence, they may require trained experts from private and NGO sectors to provide services like value chain management, market linkages, product planning, *etc.*

Private agents' capacity can be improved by strategic alliance with government institutions. Such PPP provisions can be highlighted in SREP and SEWP through ATMA mechanism according to local needs. Example: training centre in Khurda district strengthened by Paradeep Phosphate by providing training faculty and infrastructure, for farmer training activities.

Seeing the limited approaches by public sector in agricultural extension, process of privatization of extension can be taken as alternative for improving extension effectiveness. It is expected to improve by: reorienting public sector extension with limited and well-focused functions, more number of extension providers (institutional pluralism) resulting from active encouragement by the public sector to initiate, operate and expand, more private participation leading to the availability of specialised services hitherto not available from the public system, user contributions to extension leading to improved financial sustainability, and support and control by clients leading to client orientation, (Sulaiman and Sadamate2000).

Several Parameters like who initiates PPP, sharing of responsibilities and benefits, mutual trust, who takes lead, who supports, dispute resolution mechanisms, *etc.*, play an important role in sustaining PPPs (Sadamate, 2003).

5.3.2 Use Input Dealers, Entrepreneurs and Start-ups in Promoting Extension

There is adequate space for promoting Dealers, Distributors and Extension Entrepreneurs in extension services especially in motivating other non-users while promoting their own product and services. Extension through ACABCs and Agri Start Ups is also gaining ground in commercial crops. It has been observed that a Successful venture deploys 10 workers directly and about 40 indirectly, covering about 200-500 Farmers. Farm Entrepreneurs and Agri Start-ups could play major role in extension services across the production systems.

It has been observed through various studies that agri- preneur being highly motivated person could take on activities like aggregated production models, providing better forward linkages, driving market led extension strategies, better price realization to the farmers (Chandrasekra,2014).

5.4 Strengthening R&D Linkages

5.4.1 Enhanced Research Backstopping to Extension

Enhanced research backstopping to the field formations is required so as to facilitate a strong technological input in the extension processes. The scientists interface with field functionaries and with the farmers would need to be augmented. Technology options would need to be generated for the micro level farm situations. Also, validation of farmers own innovations needs to be prioritized and mainstreamed. Farmer Entrepreneurs/innovators need to be promoted in extending extension services by using them in training and capacity building activities.

5.4.2 R&D Linkages: Activating Existing Forums and Follow-up

There are various R&D linkage mechanisms available in agriculture and allied sectors at various levels as follows:

National: Pre-Kharif and Pre- Rabi Interfaces between DAC and FW and ICAR

Regional: ICAR Regional Committee Meetings

State level: Pre-Seasonal Meetings (Agro - Climate Zone Wise)

District Level: SAC of KVKs, GB/Management Committee of ATMA and District level ATMA FACs

Block level: Block Level FACs of ATMAs

Experiences in various States indicate that the performance of these forums needs lots of improvement so that they function efficiently in contents and coverage. It is also necessary that both research and development sides participate pro-actively in improving R&D agenda, making these forums dynamic and responding to the agenda as per needs of farmers. ATMA was proposed to be a platform for broad based and intergraded extension operations, the States and the districts would need to think on this proposition seriously.

5.4.3 Technology Mapping over Socio Economic Dimensions of the Farming Community

Technology options for Agro-Climatic Zones and Sub-Zones needs to be systematically worked out and integrated into programme delivery mechanism, as per specificity of each micro agro- eco situation. Socio-economic dimensions and technology options if matched to the extent possible, would address extension operations effectively. This would bring in enhanced focus in extension processes as opposed to mere TOT approach.

5.4.4 Feedback Management

Feedback collation from the farmers, field functionaries is generally weak in the extension system. So also,its systematic documentation and analysis. It needs to be strengthened at all the level. The collation of feedback (both positive and otherwise) may begin with the farmers from various categories and from various macro and micro farming situations, followed up from the field functionaries (Figure 1.5).

Systematic documentation at KVK/ATMA level and analysis at SAMETI/ATARI level would not only improve the relevance of the research, but also, it would make the extension operations flexible and far more relevant, demand driven and productive suiting to the various agroecologies and categories of farmers. It may also facilitate giving appropriate signals to the senior management, research agencies and policy makers.

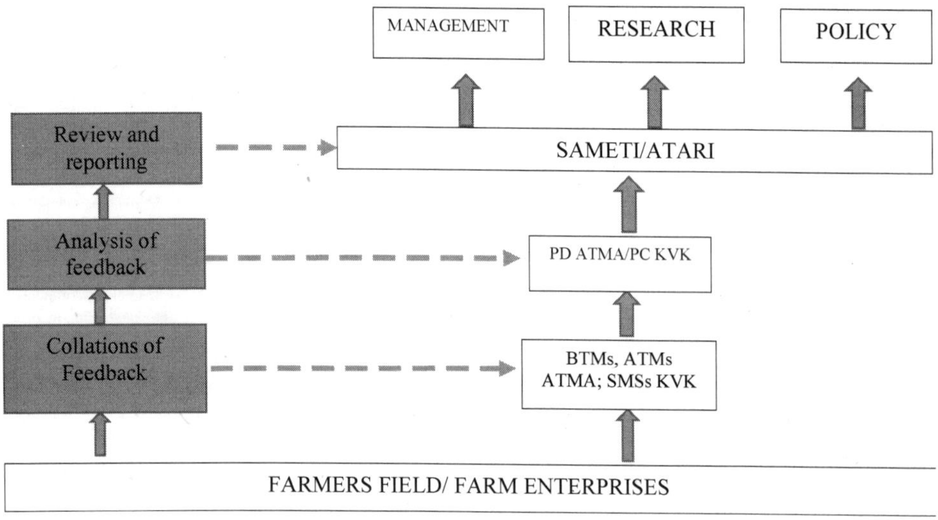

Figure 1.5: Feedback Management.

5.5 Revisiting Training and Capacity Building Efforts

5.5.1 Re-visiting Training Strategies

Training and Capacity Building is a huge task:

We have various categories of Farmers (140m), Field Functionaries (Public1.19 L+ Private 1.00L), Farmer Friends 3.3Lakhs, 660922villages,7069blocks,722 districts. Training Programmes presently operate with low coverage, lacking serious TNA exercises in vulnerable areas, inadequacies of infrastructure, experienced faculty and updated lesson plans.

Training Framework

We have a Training Framework at the National, State and District levels as: MANAGE, SAMETIs, ATARIs, EEIs, KVKs, FTCs, NGOs *etc.* Interplay of these institutions would need to be worked out systematically, sharing the output- input relationships and sharing the experiences. Capturing International experiences/ success stories in the training strategies is another strong dimension. Farm Women training needs focus on access, empowerment, CB, and market linkages. Farm Youth training focus should be on promoting youth for commercial enterprises across the production systems.

5.5.2 Capacity Building of Field Functionaries

It is estimated that over 1.19 lakh field functionaries are operating for the field extension work in agriculture and allied departments in all the Sates. This would include about 7,069 block level officers, about 722 District Agriculture Officers, about 21,207ATMs (3 per block). Further, it is assumed that there would be about 3.3 lakh Farmer Friends (one for two villages, total villages being 660922) are engaged in the field extension services from the development side. This apart we have about 1

Lakh private extension service providers in private sector in various sectors. Besides we have about 713 KVK PCs and 4,278 SMSs, 6 per KVK.

The training and capacity building of such a huge manpower is a gigantic task which could be taken up by the SAUs and ICAR institutes in a systematic way. This issue needs to be focussed adequately. The scientific institutions like SAUs and ICAR Institutes have a great scope for updating technical competence of the SMSs and middle level functionaries. The management input would need to come from the MANAGE and SAMETIs. Modalities may have to be worked out jointly by the ATARIs and SAMETIs under the guidance of Extension Division of ICAR and MANAGE. Similarly, the KVKs and FTCs of the State Departments of Agriculture could take on the training and capacity building task of the huge number of Farmer Friends, farmers/farm women and farm youth.

5.5.3 Training for Farm Woman

Farm Women play major roles in farming and their presence is increasing in all the farm operations (Agriculture, horticulture, animal husbandry, dairy, poultry, fisheries, sericulture, *etc.*) due to a variety of socio-economic factors. Male migration to towns and cities is one of the main reasons. Farm women need to be empowered by various means like grouping them into SHGs, Farm Women Groups, Farm Women Organizations, *etc.* Further, the extension services are manned by male workers, barring a few States and the development services are not reaching the women farmers at the desired level. Serious attention needed to improve the access of farm women to the extension services, both public and private and across the sectors.

There is need to orient these services in favour of favour of women farmers training and capacity building and customized training and capacity building modules need to be operated as per the needs of the farm women at a micro level. Therefore, there is need for intensive capacity building and empowerment efforts for women and women groups at every level that should include re-orienting farm services, improving their access to credit, technology, inputs, markets and gender friendly mechanization. It may call for feminization of farm extension services enabling effective women extension workers for better participation and extension delivery. Farm women including women from weaker sections need to be given adequate representation in various programme management forums at all the levels. The ongoing programmes and schemes should ensure greater gender coverage through improved participation, financial allocation and well defined extension strategies.

5.5.4 Training of Farm Youth

Farm Youth is no longer interested in primary agricultural operations. Rather, he/she is looking for opportunities in commercial agriculture, high tech horticulture, commercial dairy and fisheries *etc.* Farm Youth are found to be taking up lots of interest in farmer empowerment models like Farmer Organizations, Producer Companies, mechanized operations, processing and value added activities, *etc.* Such operations would give him/her a better comparative social scale in comparison with his/her urban counterparts and peers. Training for farm youths and subsequent follow up should be oriented accordingly. ICAR initiative like ARYA, entrepreneurship windows opened up under RKVY, ACABCs, Skill Mission, Dairy entrepreneurships, *etc.* are a few such examples, these are needed to be systematically popularized and expanded.

5.5.5 Training Co-ordination-Segregated Responsibility, the Directorates of Extension of SAUs to Lead the Process

Farmers training responsibility down the line in agricultural and allied sectors is shared by the large number of agencies and organizations. For example, at the district level the farmers are trained by the agencies like KVKs, ATMAs, SAU Outposts, FTCs, Input Support Agencies, NGOs, *etc.* Can we think of focussed and segregated training responsibility? Could Director Extension, SAUs, ATARIs, EEIs and SAMETIs take a call in respective State and address this challenge? MANAGE and Extension Division of ICAR may take an overview and guide the process. It would streamline the whole training function and deliver it better.

The Directorates of Extension of SAUs would need to play an important role in coordinating the training function in a State in collaboration with the SAMETIs. Even their role in outreach programmes is observed to be centred around the KVKs. The role of these Directorates in training management, information support, field/outreach programmes that were visible in mid-eighties would need to be revisited/reworked in the changing agricultural extension demands. They may have to focus on PPPs, promoting entrepreneurships, start-ups, incubations, R&D linkages, extension in allied areas, studies and action research, media management including social media, *etc.* Accordingly, the manpower and investments in their operations would need to be enhanced.

5.6 Extension Planning/Prioritization and Convergence

5.6.1 Extension Planning and Prioritization

Extension issues must reflect in Block Extension Plans(BEPs), at the cutting edge levels, defining role space of various ESPs

There are series of district level planning instruments available as follows:

- ✰ Comprehensive District Agricultural Plans (C-DAPs) under RKVY,
- ✰ Strategic Research and Extension Plans(SREPs) under ATMA
- ✰ District Profiles(DPs) prepared under KVK Programme
- ✰ Detailed Project Reports(DPRs) developed under Watershed Projects
- ✰ Potential Liked Credit Plans (PLPs) developed by NABARD for the districts
- ✰ District Irrigation Plans (DIPs)) developed under PMKSY.

The District/Block extension functionaries may scan these instruments for capturing the extension priorities and integrating the same in the convergence and programme delivery strategies.

5.6.2 Convergence and Operational Flexibilities

5.6.2.1 Agricultural Technology Management Agency (ATMA): Convergence Platform

Programmatic Convergence down the line is essential for wider coverage and to benefit from each other's strengths. Policies and instructions are necessary to drive it. Convergence appears to be most effective at the Block/Cluster level. Matrix mode/MOU based approach is recommended to make it happen at the field level.

There is need to revisit the contents and methodology for development of Block Action Plan (BAP) which shall invariably indicate role space for both public and private sector extension service providers, including technical inputs from KVKs and financial support from all sources. Accordingly, the Block level, Cluster level and Village level extension operations will need to be streamlined to meet emerging challenges in farming system within the respective blocks.

ATMA shall promote both Public and Private Extension Service Providers (ESPs) to facilitate coverage of extension operations in entire district, in view of manpower and resource constraints. These Extension Service Providers may include Agri-preneurs, Agri-Start-ups, Farmers Knowledge Groups, Farmer Producers Companies, Agri-business Companies, Farmer Organisations, Media Channels, Exporters, CSR Foundations, NGOs, Achiever Farmers and SHGs. A systematic data base, Competency Development Plan and their location specific involvement may be worked out as per demands of BAP. Perhaps, similar recommendation is applicable for KVK also. ATMA has to maintain directory of Private Extension Service Providers including Agri-preneurs in the district.

Convergence Matrix (CM) may be prepared which will list out scheme and component wise allocations and activities. Every proposal for funding under existing schemes or new schemes shall be accompanied by a Convergence Matrix relating to that Department with a detailed explanation about to what extent convergence has been achieved. Every authority scrutinizing and approving the programme must do convergence scrutiny to satisfy himself/herself that convergence has been brought about. Departments and institutions in the convergence matrix shall co-opt the departments and the related institutions in decision making.

The following need to be attended for convergence as per DFI Committee Report 2018.

a) Monitoring of convergence arrangements between ATMA, KVK and Line Departments may be accorded highest priority. The process may be reviewed in every SAC of KVK and ATMA GB meeting.

b) Implementation of Convergence Guidelines (17.06.2015) jointly issued by DAC and FW and DARE/ICAR may be taken up with ATMA Officials and KVK Scientists at State/District level with SAMETI/MANAGE facilitation;

c) This may pave way for inclusion of convergence module in the regular training programme of MANAGE, SAUs, ATARIs, EEIs and SAMETIs *etc.*

5.6.2.2 Convergence with other GOI Programmes

Agriculture development programmes and related programmes of other departments like rural development; Panchayati Raj, Social Development, *etc.* generally operate in vertical silos, with very little or no horizontal convergence, especially at the block level and below. Hence, field level convergence is one of the major challenges existing today across the departments and levels. The major task, therefore, would be to reorient the district and the block functionaries for the task and provide them operational flexibilities and opportunities for experience sharing across the departmental boundaries so as to obtain required coherence.

- ✰ Matrix mode schematic alignment
- ✰ Operational flexibility across department
- ✰ MoU based
- ✰ Gap filling mode

Likewise, the programme delivery is crucial in disadvantaged areas, far flung areas, tribal areas, hill and mountainous areas, desert areas, *etc.* Hence, Extension Strategies need to be re-oriented and strengthened accordingly on convergence mode. Flexibilities are required to be promoted to the field formations to respond to the emerging local extension challenges. Thus, making extension operations accountable to the local situations/farming communities. Operational flexibilities may facilitate entering local MOU based PPPs which are essential to make the extension interventions work at the local level.

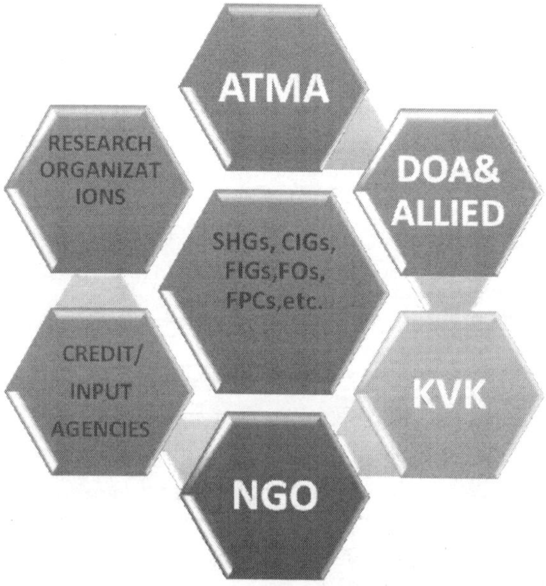

Figure 1.6: Action Research on Convergence based Extension under NASF.

5.6.3 Block Extension Plans as an instrument of Convergence

Convergence requires proper role space and resources amongst stakeholders for mutually agreed action plan, drawn from SREP. Matrix mode approach is required to be operated which would indicate the ongoing programmes and activities of the stakeholders and the gaps. The resources would need to be pooled accordingly and the roles and responsibilities defined in Block Extension Plan (BEP). The private and public extension service providers would need to agree for the area and location specific roles as per competence and presence of their system. The decentralised extension system with farm schools, farmer friends and PPPs needs to be exposed to the contemporary technological developments and timely and accurate advisory services.

Besides ATMA led field extension services at the district, block and cluster levels, it would require careful harmonization of work plans of the *Rashtriya Krishi Vikas Yojana* (RKVY), national missions, and other related State schemes and their resource allocation and schematic support to succeed convergence of activities. This will not only increase the operational resources for effectively targeting the ATMA activities but also will help national schemes to meet their objectives and make ATMA sustainable in the long-run (Babu *et al.*, 2013)

5.7 Media Management/Information Support in Agricultural Extension

Information support to Technology Dissemination in agriculture and allied sectors broadly falls in three categories namely print media, the electronic media and the other media like social media. Each having large scope to influence farmers and farm women by depicting success stories, entrepreneurial experiences, *etc.*

5.7.1 Information Sourced by the farmers

Smallholder farmers were found to rely mainly on local sources of information, such as progressive farmers (16 per cent) and input dealers (12.6 per cent), along with the radio (12.4 per cent). In case of **medium-size and large farmers**, the major sources of information in addition to those above were television (15.3 and 22.4 per cent), newspapers (10.3 and 15.9 per cent) and extension agents (9.8 and 12.4 per cent). To put it differently, **only 4.8 per cent of smallholders** viewed the extension worker as a primary source of information, as compared to 9.8 per cent of medium farmers and 12.4 per cent of large farmers (Adhiguru *et al.*, 2009). Glendenning *et al.* (2010) concluded from a review of agricultural extension in India that despite the variety of agricultural extension approaches that operate in parallel and sometimes duplicate one another the majority of farmers in India need improved access to information.

5.7.2 Media Management for Promoting Extension

- Present media coverage to allied sector is < 10-15 per cent
- Subject centric contents, programme treatment and delivery needs to be improved
- Intensive use of the following is suggested:
- Internet technology and internet platforms,

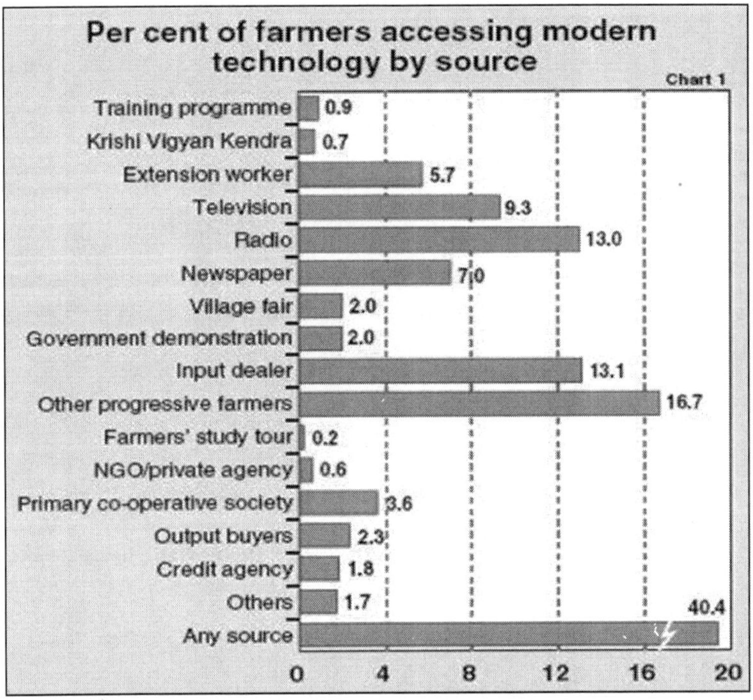

Figure 1.7: Information Sourced on Modern Technology by Farmers, 2003 NSSO Survey.

- Social media and Apps
- Champion's approach and its publicity
- Portal indicating services and linkages

5.7.3 Krishi Darshan and DD Kisan

Krishi Darshan programme is being telecast 5-6 days in a week through national (1), regional (18) and HPT/LPT transmitters (180). Recently launched DD Kisan channel provides 24-hour agriculture based information to the farming community covering research updates, extension advisories, market rates and weather updates

5.7.4 Community Radio Station (CRSs)

All India Radio through their Kisan Vani Programme broadcasts farm programmes 6 days in a week through 96 FM stations.

Community Radio Stations (CRS) provide updated agriculture information which is location specific to a group of villages. CRS can reach local population instantly through time and cost effective means. ATMA made provisions for establishment of CRS in all the districts by providing funds for infrastructure and content development for two years. However, response is discouraging. Only few CRS stations have been established. Some of the important obstacles are delay in obtaining the license, restrictions on advertisement time and challenge in creating attractive content. CRS should be the mouth piece of all the public and private developmental departments working in the coverage area. This approach provides opportunity for disseminating varieties of information from different departments. Thus, financial viability of CRS will be enhanced.

All SAUs, ICAR Institutes may consider having CRS for location specific broadcasts for dissemination of farm information.

5.7.5 Field Days and Campaigns

5.7.6 Use of Vernacular Press and Traditional Folks

5.7.7 Kisan Call Centres

A country wide common eleven-digit Toll free number 1800-180- 1551 has been allotted for Kisan Call Centre. This number is accessible through all mobile and land line telephone networks including private service providers. Replies to the farmers' queries are given in 22 local languages and the calls are attended from 6AM-10 PM on all the seven days of the week at each KCC.

The experience of rural tele-centers in the developing world shows that ICT can help in enabling rural development workers to gather, store, retrieve, adapt, localise and disseminate a broad range of information needed by rural families (Davison *et al.*, 2005). The ICTs in extension can lead to the emergence of knowledge workers that will result in the realisation of a bottom-up, demand-driven paradigm for technology generation, assessment, refinement and transfer (Meera *et al.*, 2004). It is estimated that there are 251 million internet users are in rural India which is expected to reach by 290 million by 2019 (The economics Times, 6thMarch,2019). There are a number of initiatives in operation using mobiles to communicate information directly to farmers

5.7.8 Social Media

Social media is a powerful communication tool that widens lines of communication, enhances user network, engages a large number of people quickly and has a great potential to reach out farmer's/farm women and stakeholders.

5.7.8.1 e-Platforms

Emails, What's app Groups, Face book, Blogs, Instagram, Twitter handles, Commercial apps *etc.* are also gaining importance in farming.

A study on WhatsApp enabled social media by (Chander, 2016) indicates that by the use of WhatsApp farmers are able to seek information on farm operations, clarify their doubts on plants/livestock disease symptoms and are having immediate access to market related information. In the Indian State of Karnataka, the Department of Agriculture has made it mandatory for its development officials to have a smart phone so that they could share information, messages, and circulars through WhatsApp.

WhatsApp can be used to offer context-specific information by linking farmers on one side and resource persons of different disciplines on the other interlinked through a common mediator. The role of the mediator would be to receive the queries from the farmers on one side and sending them individually to different experts. After receiving the queries from different experts, the mediator can share the pooled advice in an easily understandable form to the farmer clienteles (Thakur, 2016).

5.7.8.2 Farm Portals

These are also available for dissemination of information, e-commerce and distance learning are also promoted. Information given includes crop production and protection technologies, inputs, prices, e-commerce, extension needs, *etc.* Portals vary widely in their contents, updates, user friendliness and use of visuals.

5.7.8.3 ICTs in Extension Cycle

ICTs and Media combos should be promoted as per need of the extension cycle (Awareness/Interest/Evaluation/Trial/Innovators/Early adopters/Early Majority/Late Majority and Laggards)

ICTs are also being used to strengthen the capacity of extension officers and NGO field staff to reach farmers with timely and accurate information and, at the same time, help capture data from the field(Chander, 2016). ICT has the ability to reach farmers directly, can enable two-way information sharing processes, has greater storage capacity, is faster, and can increase market efficiency by addressing information gaps and blockages (Chapman and Slaymaker 2002).

e-Chaupal-The e-Choupal initiative has had positive effect on the incomes of participating farmers, as the system has brought efficiency to the supply chain by removing intermediaries and reducing transaction costs (Bowonder, Gupta, and Singh 2007).

ICT has lots of scope in future extension strategies, especially at the awareness and interest stages of adoption process. However, content development, validity and delivery in respect of print and electronic media is crucial. ICT is becoming very popular day by day, however, the pertinent question is how to increase its access and reach to the large number of farmers. Mobile applications, internet access, development of apps and their intensive use in extension need to be expanded.

5.7.8.4 Common Service Centres (CSCs)

There are 1.57 lakhs Common Service Centres located at village level providing citizen services to villagers. These are operated by village level entrepreneurs. The CSCs possess minimum IT infrastructure to provide IT enabled services to people on payment basis. CSC is great national resource available at village level, managed by entrepreneur which could be used for providing agricultural extension services by value addition to the concept. For this transformation, CSC entrepreneur has to be trained in basic agriculture and should be made aware about information sources in agricultural extension. With this empowerment, CSC can serve agricultural extension related services.

5.7.8.5 Farmer/Farm Women Knowledge Groups (FKGs/FWKGs)

There is need to organize Farmer Knowledge Groups (FKGs) and Farm Women

Knowledge Group (FWKGs) in the villages facilitating/improving arrangements for knowledge interpretation and knowledge sharing.

5.8 Addressing Extension Process Equilibrium and Future Research In Extension

5.8.1 Sequential Extension Process Equilibrium

Future Extension strategies must address extension needs of areas having high yield potentials but performing at lower levels. Extension strategies must follow sequential extension process equilibrium as: Technology Validation (TV), working out technology Options (TOs), Technology Dissemination, Participatory Processes and obtaining feedback for further improvement. There is need to organize large number of FFSs/FOs and FPCs in potential areas. Extension functionaries must identify potential areas for export opportunities (for progressive farmers) and involve processing and export industries in such assessments. Promotion of forward linkages through appropriate extension interventions and enhancing outreach to small and marginal farmers would be essential. Hence, focus on social dynamics would be required.

5.8.2 Future Extension Strategies should focus on Crucial Technologies and Concerns

Field extension functionaries from agriculture and line departments would need to be adequately oriented in promotion of crucial technologies like Climate Change, especially adaptation strategies in the context of increasing vulnerabilities. The approach would benefit the farmers in judicious and timely input use enhancing farm incomes.

5.8.3 Future Extension Strategies Must Draw Strengths From

The future extension strategies must draw strengths from international extension models and private sector experiences: learning from companies, corporate sectors, multi-nationals, Non- Government Agencies, Entrepreneurs, Farmer Professionals and Para professionals and Outreach Programmes of Research Organizations. This would enable extension contents and delivery move on improved and dynamic mode as per need of the local situation.

5.8.4 Refocus Research in Extension

Research in Extension is a crucial area but not attended to adequately to look into the efficacy of different Extension approaches being followed in various states. Strong extension research input is required to strengthen the basic knowledge in extension discipline (NAAS,2015). MANAGE, SAMETIs, ATARIs and Directorates of Extension of the SAUs would need to have a strong extension research window. They may deliberate and define broader Extension research areas annually as applicable (*e.g.*: case studies, comparative analysis of extension approaches, media combination for effective extension, training strategies, role combinations of extension service providers, extension priority setting, *etc.*). The specificities should be left with the local institutions

5.9 Revisiting Agricultural Extension Syllabus

5.9.1 Dynamic Demands on Extension Syllabus

Competent extension professionals are the assets of agricultural extension services. Diverse and dynamic agricultural systems, advancing science and technologies, changing socio demographics, increasing globalization and growing competition for resources, demand that agricultural extension professionals be proficient in the technical aspects of their areas of expertise as well as in the processes and delivery of the services. In other words, the need and demand for extension professionals to demonstrate a higher level of professionalism in their services are growing.

5.9.2 Crucial Recommendations on Revision of Extension Syllabus

a. Revisiting the existing syllabus of Agricultural Extension Education with the following recommendations is suggested (Sadamate and Padaria, 2018)

b. Orienting the course syllabus to the emerging challenges in agriculture and Agricultural Extension

c. Fundamentals of Extension should be continuously enriched with lessons/ experiences of contemporary Extension Reforms

d. Customization of management principles with suitable cases of extension organizations and approaches

e. Ensuring relevance of course content to present and future scenarios in agricultural extension

f. Extension Education for entrepreneurship

g. Enabling ICT mediated extension system

h. Elaborate user consultations involving experts from diverse user groups like academicians, practitioners, industries, business organizations, farmers and change agencies to provide input to ICAR/Deans Committee.

5.9.3 Emphasis upon Skill Building through Suitable Courses and Practicum

(i) *Social Skills*

- ☆ Situation and gap analysis: rural and urban systems, gender and farm youth
- ☆ Communication skills
- ☆ Rapport building with multi stakeholders.

(ii) *Management Skills*

- ☆ Need assessment
- ☆ Programme planning, execution, evaluation
- ☆ Mobilization and group dynamics
- ☆ Team building
- ☆ Decision making

- ☆ Conflict management
- ☆ Empowerment of farmers and disadvantaged groups
- ☆ Business management skill

(iii) Technical Skills

- ☆ Extension methods and interventions
- ☆ Livelihood analysis and promotion
- ☆ Entrepreneurship development
- ☆ ICTs
- ☆ Data collection, analysis, report writing
- ☆ Participatory methods
- ☆ Stakeholder analysis and their interplay
- ☆ Market assessment and Value chain development
- ☆ Contemporary issues - gender, climate change, sustainability issues
- ☆ Risk analysis

5.9.4 In addition to above, specialized skills in following areas should be developed among PG students:

Agro ecosystem Analysis

- ☆ Project formulation- feasibility, revenue models
- ☆ Programme planning
- ☆ Programme execution
- ☆ Programme evaluation
- ☆ Impact assessment

Social Research Methodology

- ☆ Market led extension
- ☆ Media, communication, ICTs content development, instructional technologies
- ☆ Information technology
- ☆ Policy analysis

6. Agricultural Extension Services in the Context of SDGs

6.1 Organized Two Conferences

April 22-24, 2017 at Hyderabad, India and (ii) May, 10-12th, 2018 at Kandy, Sri Lanka

6.2 Stakeholders

National/International Extension Experts, NGOs, Private Extension

Service Providers, Scientists from SAU and Field Extension Systems, Extension Entrepreneurs, Innovative Farmers/Farm Women, Extension Academia and Research Scholars

6.3 Guidance by

Dr.Daniel Gustafson, DDG FAO and Dr.A.K.Singh, DDGE ICAR.

6.3 Extension Services were Found to be Directly Influencing the following SDGs (Gustafson, 2017)

- ☆ Poverty Alleviation (SDG-1)
- ☆ A Food and Nutritional Security(SDG-2)
- ☆ Education and Learning (SDG-4)
- ☆ Gender Equality and Empo. of Women (SDG-5)
- ☆ Water and Sanitation Management (SDG-6)
- ☆ Climate Change (SDG-13)
- ☆ Eco-system Sustainability (SDG-15)
- ☆ Peace and Well being (of Farming Community) (SDG-17)

7. Doubling Farmer Income (DFI) Committee Report – Observations on Extension

Honourable prime Minister of India Sri Narendra Modi envisioned 7-point strategy to double the farmers' income which laid emphasis upon securing per drop more crop through *Pradhan Mantri Krishi Sichayee Yojana*; soil health management by ensuring Soil Health Cards to all farmers; promotion of cold chains and value addition; ensuring effective market through National Agriculture Market and e-mandis; streamlining Crop Insurance through *Pradhan Mantri Fasal Beema Yojana;* promotion of ancillary activities like dairy, apiary, fisheries; promotion of organic farming; and strengthening of information system (DD Kisan).

The eminent agricultural scientist Dr. M.S. Swaminathan (Farmer Commission, 2006) suggested that doubling income is feasible by bridging the large gap between potential and actual yield per hectare. He has suggested pragmatic strategy which includes enhancing small farm productivity and stability of production through soil health care, water harvesting and management, choice of appropriate technology and inputs, credit and insurance and finally opportunities for remunerative and assured marketing; focusing upon the knowledge, skill, credit and land ownership empowerment of women farmers; inclusion of high value crops, horticulture, animal husbandry, agro-forestry in the farming systems; promotion of commercial use of the whole biomass of the crop; and fixing the procurement price at cost C2 (All cost (cash and kind) incurred on production with rental value of owned land, rent for leased-in land Cost and imputed value of family labour and interest on own assets) plus 50 per cent to enable small farmers enough surplus.

Observations of Doubling Farmer Income (DFI) Committee-2018 Chaired by Dr Ashok Dalwai

7.1 Rationale

Production performance alone is not a sufficient condition for farmers' welfare. The difference between the farm gate prices and prices on the finished/processed products are very high and the farmers share in the final price is very insignificant, Further, the farmers, specially the small and marginal ones are challenged by vulnerable climatic situations and inadequate risk cover. Size of land holding is getting smaller and smaller and the economic viability of small and marginal farms has not been able to sustain the total family size for productive gain. Farm income of small and marginal farms as compared to the town and urban areas is falling far behind this is resulting into disengagement of farm youths in agriculture and allied areas. Farm youth appears not interested in primary farming but seems to be interested in secondary agriculture. Resultantly, there is a serious, mismatch between production and income realization- hence there is need to make serious efforts to enhance the income of the farmers

7.2 Growth Drivers Identified by DFI Committee

- ☆ Increasing production and productivity across the production systems
- ☆ Effective Input Use efficiency and Programme delivery
- ☆ Reduction of Post-Harvest Losses
- ☆ Crop diversification and Value addition
- ☆ Reforms in Agricultural Marketing
- ☆ Risk Management
- ☆ Focus on Allied Sectors and Non-farm income

7.3 DFI Committee's Relevant Extension Recommendations

Redefined Agricultural Extension as:

The System of empowering farmers for sharing knowledge, technology, skills and farm management practices, across agricultural production systems and sub sectors, all along the value chain, enabling the farmers to realize the higher net income from their enterprises, on sustainable basis.

Major recommendations as they pertained to agricultural extension services are as follows:

- ☆ Providing Synergetic Role Space to Private Extension Services
- ☆ Rejuvenating the ATMA
- ☆ Strengthening KVKs for effective backstopping ATMAs
- ☆ Streamlining Agricultural Extension System, especially interplay between the systems
- ☆ Strengthening Downstream Institutional Linkages
- ☆ Addressing Human Resources Issues in Agricultural Extension

- ☆ Enabling ICT driven Extension Services
- ☆ Focus on Gender Concerns
- ☆ Revisiting Agricultural Extension Curriculum in Agriculture Universities, and making a good mix of Academic and practical, entrepreneurs, start-ups, commercial propositions.

8. Conclusion

1. The technology dissemination services, both public and private, across the production sectors, have played crucial role in achieving impressive performance of agriculture and allied sectors. Despite record achievements the agriculture sector as a whole faces formidable challenges. These include: sliding down GDP contribution of farm sector, shrinking land holdings, adverse impact of climate change, regional imbalances, emergence of farm distress though sporadic, stagnant productivity in some crops/areas, increased production costs, dwindling natural resources, market uncertainties and efficacy of delivery systems.
2. Agricultural Extension services have travelled through series of programmes like: Grow More Food campaign, National Extension Service, IADP, IAAP, NDS, FTCs, SFDA, KVKs, T and V System, NAEP, ATMAs, KCC, enhanced network of Training institutions, *etc.* Parallel, the private sector extension service providers also widened their base over a period and included: NGOs, Input Support Providers, Entrepreneurs, Agribusinesses, Cooperatives, ICT based services, Mass/Social Media interventions, *etc.* leading to pluralism in extension operations.
3. Significant learning and lessons emerged, focused on much needed extension reforms like: (i) Farmer empowerment through various modes,(ii) Enhancing outreach of extension models in operations, huge task, bringing in agri business and marketing opportunities, (iii) Focus on extension coverage in allied sectors, (iv) Integrating private sector efforts, (v) Strengthening R&D linkages at different levels,(vi) Revisiting training strategies, (vii)Extension planning and prioritizations, (viii) Convergence of extension efforts through block extension plans, (ix) Media management and (x) Revisiting agricultural extension syllabus.
4. Role of Extension Services has been found to be crucial in addressing the SDGs (Gustafson, 2017) and in Doubling Farmer Income Strategies (Dalwai, 2018)
5. The Extension Delivery is a very complex system dealing with farm technologies on one hand and the social issues on the other. The basic sciences have laboratory back up to generate the body of knowledge. However, in extension services we have farmer's socio economic background and his/her fields as live laboratories. Thus, social dynamics plays an important role in determining the performance of the extension services.

6. In a given situation, you may have technologies, may have schemes and programmes and investments too. However, in the absence of streamlined and reformed extension services the programme delivery to the targeted clientele may not go efficiently, the development efforts would still be inadequate to produce desired impacts.
7. Hence, sincere arguments for (i) strong extension reforms both in public and private systems, (ii) on the lines proposed, for (iii) efficient technology dissemination resulting in (iv) higher adoption, integrating marketing/ agri business opportunities and enhancing farmer incomes.

REFERENCES

Adhiguru, P., Birthal, P. S., and Kumar, Ganesh P. 2009. Strengthening Pluralistic Agricultural Information Delivery Systems in India. *Agricultural Economics Research Review*, 22(1): 71–9.

AESA. 2019. Extension and Advisory Services in Scaling up Climate Smart Agriculture in South Asia, Policy Brief.

Babu, S. C., Joshi, P. K., Glendenning, C. J., Kwadwo, A. O. and Rasheed, S.V. 2013. The State of Agricultural Extension Reforms in India: Strategic Priorities and Policy Options. *Agricultural Economics Research Review*, 26(2): 159-172.

Bowonder, B., Vinay, G., and Amit, S. 2007. Developing a Rural Market e-hub, The case study of eChoupal experience of ITC, Socio-Economic Report Planning Commission Study-ICT4. [*Retrieved from http://planningcommission.nic.in/reports/ sereport/ser/stdyict/4 e-choupal, 20*]

Chander, M. 2016. WhatsApp in Agriculture? Blog/Global Forum on Agriculture (GFAR) [*Retrieved from https://blog.gfar.net/2016/07/14/whatsapp-in-agriculture/*]

Chapman, R. and Slaymaker, T. 2002. ICTs and rural development: review of the literature, current interventions and opportunities for action. Overseas Development Institute London.

DAC and FW 2018. Annual Report 2017-18. Ministry of Agriculture and Farmers Welfare, GoI, New Delhi

DAHD and F 2019. Annual Report 2018-19. Ministry of Animal husbandry and Fisheries. GoI, New Delhi

Davison, R.M., Harris, R.W., Qureshi, S., Vogel, D.R. and G.H. de Vreede, (Eds.). 2005. Information Systems in Developing Countries: Theory and Practice, City University of Hong Kong Press, 289pp.

DES 2018. Agricultural Statistics at a Glance. Directorate of Economics and Statistics, Ministry of Agriculture and Farmers Welfare, GoI, New Delhi

DES 2018. Horticultural Statistics at a Glance. Directorate of Economics and Statistics, Ministry of Agriculture and Farmers Welfare, GoI, New Delhi

DES 2018. Third Advance Estimates of Production of Food Grains. Ministry of Agriculture and Farmers Welfare, GoI, New Delhi

MoA and FW 2018. Doubling Farmers Income Committee Report. Ministry of Agriculture and Farmers Welfare, GoI, New Delhi

Glendenning, J.C., Babu, Suresh, and Asenso-Okyere, K. 2010. Review of Agricultural Extension in India: Are Farmers Information Needs Being Met? Discussion Paper 01048, International Food Policy Research Institute (IFPRI), Washington D.C.

Gustafson, D. 2017. Agricultural Extension Services In The Context of SDGs, Advances In Agricultural Extensions, PRDIS, Hydrabad.

IFPRI 2018. Global Hunger Index Report.

Kumar, R., Singh, B.P., and Kaswan, S. 2012. Production Driven to Market Driven Extension Approach. *Indian Research Journal of Extension Education*, Special Issue 2: 125.

Meera, Shaik N., A. Jhamtani, and D.U.M. Rao. 2004. Information and Communication Technology in Agricultural Development: A Comparative analysis of three projects from India. AgREN Network Paper No.135, ODI, January 2004.20p.

Ministry of Statistics and Programme Implementation 2019. Annual Report 2018-19.

Murray, E.V. 2008. Producer Company Model-Current Status and Future Outlook: Opportunities for Bank Finance, Financing Agriculture, pp18.

NABARD 2020. Farmer Producers' Organizations (FPOs): Status, Issues and Suggested Policy Reforms. National Paper - PLP **2020**-21.

Sadamate, V.V. and R.N. Padaria. 2018. Policy Reforms in Agricultural Extension Education: Indian Perspective. Proceedings of IAUA Golden Jubilee International Conference on Agricultural Education- Sharing Global Experiences organized during November 23-25, 2018 in New Delhi.

Priyadarshni, P., Padaria, R.N., Burman, R.R. and Kumbhare, N.V. 2019. Enhancement Of Diffusion and Adoption of Local Knowledge through Social Learning: A Case Study of White Fly Management in Cotton in Haryana, *Indian Journal of Traditional Knowledge* vol 18(1), pp 169-177.

Sadamate, V.V. 2003. PPPs in Agricultural Extension Management, Fulbright Post Doc Research Study, MSU, USA.

Sulaiman, R. and van den Ban, A.W. 2003. Funding and Delivering Agricultural Extension in India. *Journal of International Agricultural and Extension Education*, 10(1): 21-30.

Sulaiman, V. and Sadamate, V. V. 2000. Privatising agricultural extension in India, NCAP, Policy Paper.

Thakur, D. 2016. An Expert-backed WhatsApp group that works for Farmers, Global Forum on Agriculture (GFAR) [*Retrieved from bloghttps://blog.gfar.net/2016/09/12/an-expert-backedwhatsapp-group-that-works-for-farmers/*]

Chapter 2

Reforming Agricultural Education for Strengthening Competencies of Extension Students

S.K. Kashyap[1] *and Girijesh Singh Mahra*[2]

[1]*Professor and Head, Department of Agricultural Communication, GBPUAT, Pantnagar, Uttarakhand*

[2]*Scientist, Division of Agricultural Extension, Indian Agricultural Research Institute, New Delhi*

Introduction

Today agricultural research and teaching is being practiced by teachers and students in more than 61 State Agricultural Universities (SAUs), 5 Deemed to be Universities (DUs), 3 Central Agricultural Universities (CAUs) and 4 Central Universities (CUs) with Agriculture Faculty. Furthermore, research is being carried out in more than 100 ICAR research institutes and State Departments. Government has established separate institutes like MANAGE and NIRD Hyderabad, under different ministries to carry out extension research and services. Under ICAR, there is separate 'Education Division' which is headed by the Deputy Director General (Education) and five Assistant Directors General namely, ADG (HRD-I), ADG (HRD-II), ADG (Education Planning and Development), ADG (Accreditation), and ADG (Home Science). The Education Division of the ICAR since 1995 onwards conducts the All India Entrance Examination (AIEEA) for admission to 15 per cent of the seats reserved in SAUs, CUs (*i.e.*, AMU, BHU, Biswa Bharti, Nagaland University, and CAU, Imphal) for UG degree (BSc, BTech., B.F. Sc, *etc.*) in agricultural sciences, namely, horticulture, agriculture engineering, dairy technology, forestry, food sciences, fishery sciences, home sciences, sericulture, and agriculture marketing and cooperation. The Education Division of the Council, towards addressing the veritable asymmetries, undertakes planning, development, coordination and quality assurance in higher agricultural education in the country and, thus, strives for maintaining and upgrading quality and relevance of higher agricultural education thought partnership and efforts of the components of the ICAR-Agricultural Universities (AUs) System comprising 61 State Agricultural Universities (SAUs), 5 Deemed to be Universities (DUs), 3 Central Agricultural Universities (CAUs) and 4 Central Universities (CUs) with Agriculture Faculty. Quality assurance in higher agricultural education in the country has been pursued through policy support, accreditation, framing of minimum standards for higher agricultural education, academic regulation, personnel policies, review of course curricula and delivery systems, development support for creating/strengthening infrastructure and facilities, improvement of faculty competence and admission of students thought All India competitions. In addition to above activities, the Education Division of the ICAR also looks into the various important activities for the agriculture education in the country such as accreditation for quality assurance, global competitiveness in human resource development, distance education for reaching to the un-reached people, fellowship and scholarships as a tool for human resource development, women technological empowerment, faculty competence improvement, and networking for access to information Overall it can be concluded as 'Extension Education' has established itself as profession and discipline with three important components of teaching, research and service.

Challenges before Agricultural Education

1. Adequacy of Agricultural Education Course Curriculum in Indian Agricultural Universities

NAAS (2005) noted the important challenges in formal agricultural education in India which need to be addressed immediately: *keeping pace with the fast changing scientific and technical improvements, curriculum up gradation, skill based education, balance between classroom teaching and practical sessions and experiential learning based on actual work-based experience.* Analysis of the present agricultural education system indicates that despite finalization of academic regulations for *UG and PG programmes after Third Deans Committee report and initiatives under the Agricultural Human Resources Development Project (AHRDP) and later on through IV Deans committee and the National Core Group, wide variations exist in quality and performance, Universities still suffer from poor governance*. The system as a whole has not taken full advantage of modern tools of management for efficient governance. The faculty in SAUs has dwindled with majority chunk of the posts remaining vacant due to financial crunch. Besides curriculum and curriculum delivery have not been changed keeping in view global technology development. **There is also no link of curriculum to employment in private agribusiness and processing industries and meeting the demands of extension.** According to an estimate (IAMR, 2001), 43 per cent of the graduates and 23 per cent of postgraduates (M.Sc.) find difficulty in accessing gainful employment. Hence, in order to ward off rising unemployment because of excessive dependence on public sector jobs, there is need to develop graduates/postgraduates who create their own work and provide work to others also. In this pursuit, futuristic Agricultural Education will need to be enriched with real life subjects and their delivery through project work rather than through mere classroom lecture alone. Although the recently revised curricula and syllabi of UG and PG programmes have to some extent addressed these issues, still there is need for reorienting them to address stakeholders concerns and needs. **Rasheed Sulalman V. (1996)** studied the existing M.Sc. curriculum in from 12 agricultural universities in the country and concluded constraints as follows

- ☆ **Repetition of the UG curricula:** Many of the topics that appear in the M.Sc curricula are repetitions of what is already covered or should have been ideally covered in the UG curriculum in Agriculture.
- ☆ **Lack of focus:** P.G. students in present curriculum would be learning the fundamentals (mostly theory) without developing expertise in any of these fields. Lack of a clear focus in the M.Sc curricula could be due to the static nature of the curriculum and also due to lack of a clear perception on the nature of skills needed to tackle the emerging professional needs and challenges
- ☆ **Absence of specialization:** The present curricula do not provide scope for any specialisation. Need for specialisation within the discipline was realised at least 15 years back but nothing was done to achieve it.

2. Lack of Teaching Effectiveness

The aim of education today is to prepare the learners for life and all round development of human personality **(Kumar and Hansra, 2000).** Upon teachers stands the whole structure of education, he is the key to quality of education **(Aggarwal, 1967).** Of all the factors which influence the quality of education and its contribution to national development, the competence and character of teachers are undoubtedly the most significant **(Education Commission, 1964-66).** Teaching skills are set of teaching behaviors which are especially effective in bringing about desired change in pupils **(Kumar, 1998). Dosajh (1977)** reported that skills involved in teaching are introduction of topic, questioning, dealing with answers, stimulus variations, use of teaching aid, non verbal cues, reinforcement, use of illustrations, lecturing, explanation, group discussion management, planned repition, teacher liveliness and closure. **Sharma *et al.* (1991)** compared usage of teaching methods at agricultural universities and concluded that lecturing is the most prominent teaching methods with no or little practical exposure. **Tripathi and Singh (1997)** found that most of the teachers in agricultural universities are lacking necessary communication skills and hardly update their teaching materials. **Mahra *et al.* (2014)** identified constraints of formal agricultural education in India as '*Skills acquired in agricultural universities/ institutes mismatch with skills required in job*', '*Agricultural universities/institutes are unable to impart essential employable skills among agriculture students*' and ***'Agricultural universities/institutes generally not focus on imparting practical knowledge rather believe in theories and lectures'***.

3. Employability of Agricultural Graduates

The agricultural graduates up to the end of 1980 have found employment mainly in the State Departments of Agriculture. With the nineties, opportunities in these public sector departments dwindled. The agricultural graduates are now competing for limited job opportunities that are presently emerging in development NGOs and agro-input and marketing companies. But now these organizations are looking for graduates or post-graduates with more rural management, communication and marketing skills. Many of these skills are generally not present in the graduates or post-graduates supplied by the agricultural universities **(Sulaiman and Van den Ban, 2000).** With the phasing out of T and V funding, many of the State Departments of Agriculture from the late 80's started giving more emphasis to group approaches in transfer of technology. These became much more important when these departments started implementing programmes such as Watershed Development and Integrated Pest Management. Increasing emphasis is now given to identification of farmers' needs at the grass roots level, bottom up planning, development of extension plans, *etc.* Many of these social science skills are lacking in the agricultural graduates and even post-graduates working in these departments. Lack of a practical orientation to the curricula based on the felt needs of the potential employers has been the main reason for this situation **(Sulaiman and Van den Ban, 2000).**

Reforming Agricultural Education for Strengthening Competencies

1. Assessment of Farmers Need

Farmers presently require a different kind of support from agricultural experts than they received in the past. The conventional model (one way, top down) of Transfer of Technology is no more valid in most of the areas and enterprises. To become relevant, agricultural education have to integrate knowledge from more diverse fields to meet the needs of their clients and have to work with more diverse organizations.

2. Knowing Job Role and Competencies Required

The changing job markets for agricultural graduates different skill set for example in extension, market demands professional managers (not workers), media planners/communication specialists (not audio-visual operators or communicators) and HRD managers/training specialists (not training organizers). Agricultural professors and placement cell should listen carefully to (potential) employers on the criteria they use in selecting new staff members.

3. Matching Teaching Methodologies with Competencies Required

Sulaiman (1996)identified urgent need of three specialization and competency exclusively needed for each area of specialization to be developed under P.G. students of Extension Education, and hence the extension teaching methods also need a reorientation as follows. Thus competency requirements and ways to strengthen them can be summarized in Table 2.1.

Table 2.1: Competency Requirements for Extension Students and Teaching Methodologies to Strengthen them

<table>
<tr><th>Area of Specialization</th><th>Competency Required</th><th>Teaching Methodologies Required (beside lecturing)</th></tr>
<tr><td>Extension management</td><td>Building and sustaining organisations; need assessment, paradigm shifts in rural development; planning and implementing development programmes; optimum utilisation of human and technological resources in programme implementation; technology evaluation, management techniques in planning; monitoring; and evaluating projects, using the results of monitoring to modify and change programme directions, emerging institutional changes in transfer of technology, participatory extension methodologies, information management and MIS, research in extension management</td><td rowspan="3">1. Experiential Learning,
2. problem solving approach,
3. Case studies,
4. Workshop teaching,
5. Rural assignments,
6. Action research,
7. Field investigation method,
8. Project method,
9. Demonstration method</td></tr>
<tr><td>Development communication</td><td>Communication technology and media materials, media planning, production of different projected and non-projected media, participatory message design and development, evaluation of communication media for effectiveness (outreach and expected results) and cost, techniques of photography and its application in scientific presentation and extension, developing agricultural communication projects, designing visuals for print and TV/video media, agricultural journalism development communication research areas, approaches and methodologies</td></tr>
<tr><td>Human resource management/ training</td><td>Strategies in Human Resource Development (HRD); organisational behavior and group dynamics; types of extension organisations; strategies for organisational effectiveness in extension organisations; entrepreneurship development; preparation of HRD plans for extension organisations, experiential learning methods training - needs assessment, methods, and evaluation; designing effective training.</td></tr>
</table>

4. Improving Present Teaching Infrastructure at Indian Agricultural Universities

During the early phase of SAUs' establishment, funding from Centre and State was fairly adequate for development of infrastructure including laboratory facilities, equipment, libraries and research farms. Upto VI Plan, almost 33 percent of the ICAR budget was devoted to strengthening of agricultural education in the country and this is the major reason that most of the universities established during that period have excellent infrastructure, which is largely contributed by the ICAR. The share of agricultural education in ICAR budget was 8.9 per cent in VIII Plan, which increased to 14 per cent in X Plan and further to 21.5 per cent in XI Plan. In absolute figures, it has increased from Rs. 224.69 crores in the IX Plan to Rs. 2900.00

crores in the XII Plan. However, with time, the number of universities and their constituent colleges and departments has increased but the budgetary provisions could not be increased commensurately.

Conclusion

The quality of agricultural education is governed interactively by faculty, infrastructure and curricula. All over the globe, Different universities are molding their Extension teaching pedagogy. **Fifth Dean Committee report of Agricultural Education Division, Indian Council of Agricultural Research** noted that there has been a decline in quality of students and education due to obsolete and inadequate equipment, laboratory, farm and library facilities, leading to knowledge deficit all along the value-chain. Although, new buildings particularly, class rooms and examination halls have been constructed but still they are inadequate to cater to the present day requirements. There is urgent need to equip agricultural universities with adequate infrastructure for experiential learning.

In India several concerns have been expressed for declining quality of agricultural graduates. Major concern are depleted faculty strength, lack of manpower in frontier areas, inadequate hands-on skills and lack of research experience **(Kumar *et al.*, 2014)**. The SAUs also have a serious problem of inbreeding as nearly 51 per cent of faculty members have their all degrees from the same university in which they are teaching **(Ayyappan and Arunachalam, 2014). Anupam Varma (2014)** studied the agricultural student's perception about teaching effectiveness from 14 agricultural universities and concluded following immediate changes

Reorienting Agricultural Education: Teaching Learning Environment

- ☆ Present curriculum needs complete change: need to be updated with emerging subjects and a focused specialization. It has to be matched with job market demands
- ☆ Teaching methods needs to be reoriented for achieving teaching effectiveness: Case studies, internship, Experiential learning, group discussions should be introduced. Term papers should be introduced at PG teaching. ICT should be harnessed effectively; a proper mix of blackboard and power point presentation is required. Like IIMs teacher must provide background reading materials before class and class should be like discussion
- ☆ Teachers should have exclusive responsibility to teach, and no research and teachers must update their knowledge with abroad trainings frequently. Student exchange and Faculty exchange program with foreign universities is highly required. Teachers need to update regularly; many teachers refer to very few books and teach with old notes
- ☆ Introduce feed back after each class, to judge the knowledge gained. A strong feedback mechanism for teachers is highly required
- ☆ Practicals are highly neglected in most of the SAUs. It should be taken by experts with illustrated compendia. Teachers should create interest in practical through creative exercises, like preparation of models. Practical

classes should be held in proper laboratories and fields. Agricultural PG students should perform field experiments in actual farmer's fields and in examinations students should be tested for understanding the basic concepts behind experiments. Develop practical exercises directed at solving problems; taking local problems.

- Agricultural teachers should have contacts with a large number of farmers, so that one student is assigned interaction with one farmer. Extension PG students should be attach the students to a village

The theme of the XI Agricultural Science Congress of the National Academy of Agricultural Sciences in Bhubaneswar, 7-9 February, 2013 was Transforming Agricultural Education for Reshaping India's Future. The Congress emphasized the need for creation of a world-class agricultural system **(Singh, 2014)** which will be only possible if we adopt World class teaching pedagogy in agricultural education. The ICAR Vision 2030, also emphasized the need to strengthen and streamline higher agricultural education system to enhance quality of human resource in agri-supply chain to meet the future challenges through cutting-edge science and technology

REFERENCES

Chowdhary, K. 1973. Comment on Working Paper 1 – Social Sciences in Agricultural Education. In: Social Sciences in Professional Education (Agriculture, Engineering and Medical), Proceedings of the conference on the Role of Social Science in Professional Education. New Delhi, Indian Council of Social Sciences Research.

Dosjah, N.L. 1977. Modification of teacher behavior through microteaching, Sterling Publishers Pvt. Ltd. New Delhi.

Kumar, R. 1998. Students Perceptions on classroom teaching behavior of teachers –A study

Kumar, A., De, D. and Tamminana, S.K. 2014. Quality education for quality production in agriculture, *Agriculture Update* 9(4): 578-583.

Rasheed Sulaiman V and Anne W. van den Ban 2000. Reorienting Agricultural Extension Curricula in India. *Journal of Agricultural Education and Extension,* 7(2): 69-78.

Verma, A. 2014. Agricultural Education in India: Imaging Possibilities to Meet Challenges in the Changing World, National Agricultural Education Day Lecture IARI, New Delhi, November 13, 2014.

Chapter 3

Farmers' Empowerment Through Farmer Producers Organizations: A Community Initiative for Entrepreneurship

G. Eswarappa[1], *V. Veerabhadraiah*[2] *and S. Prabhu Kumar*[3]

[1]*Former Director of Extension, UAS, Bengaluru and Vice President – ISEE (South Zone) and Vice President, Institution of Agricultural Technologists, Bengaluru*

[2]*Former Director of Extension, UAS, Bengaluru and President, Institution of Agricultural Technologists, Bengaluru*

[3]*Former Zonal Project Director, ICAR- Zonal Project Directorate, Zone I Ludhiana and Zone VIII Bengaluru*

- ☆ The Farmers Producers Organization (FPO) is a collectivization of producers, especially small and marginal farmers, into producer organizations and emerges as one of the most effective pathways to address the many challenges of agriculture most importantly, improved access to investments, technology and inputs and markets. Recently it is the most appropriate institutional form around the world to mobilize farmers and build their capacity to collectively leverage their production and marketing strength.
- ☆ FPO structure: Cluster identification → Baseline assessment and diagnostic study → Farmer Interest Group (FIG) (15-20 farmer) Min 50 FIGs consisting

1000 farmers → FPO → Executive Body (2 farmers/FIG) → Board of Directors → CEO.

- The Department of Horticulture, Government of Karnataka has initiated formation of FPOs from 2014-15. So far 90 FPOs have been registered under Companies act 2013 covering all the 30 districts of the state. Roughly 2-3 FPOs were established per district. All the FPOs were established on commodity crop based (2-3 major horticulture crops). The FPO concept has been effectively implemented in collaboration with Small Farmers Agribusiness Consortium (SFAC, New Delhi), a nodal agency.

Introduction

Karnataka State is in the forefront with regard to the development of horticulture in the entire country. Horticulture area is 19.39lakh ha. (Ranks 3rd in India). Horticulture production: 196.66 million tonnes (Ranks 7th in India). Horticulture productivity: 28.52 metric tonnes per ha. Horticulture sector contributes Rs. 36,103 crore *i.e.,* 53 per cent of the production under the combined agricultural sector in the State. Around 22.50 lakh farm families are engaged/dependent on the Horticulture sector in the State.

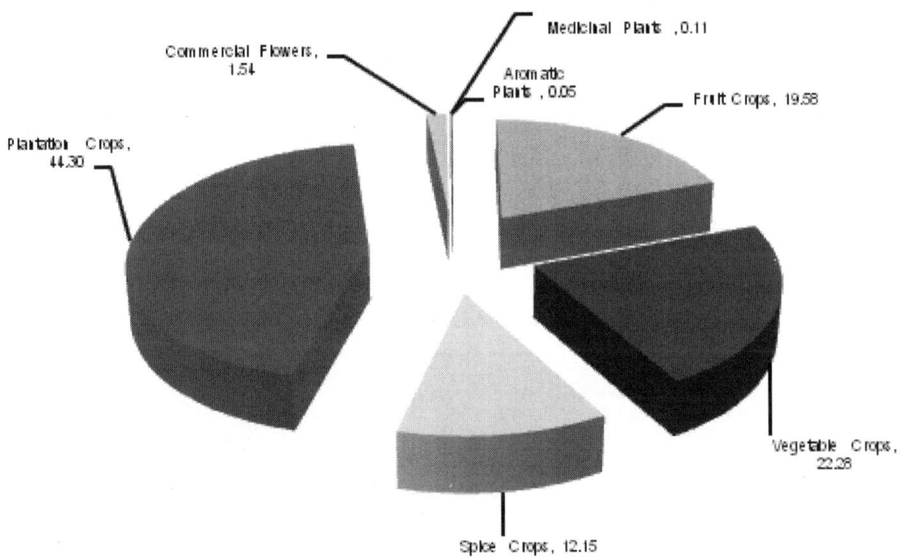

Farmer Producer Organization (FPO)

Objectives

To promote economically viable self governing, democratic farmer producer organizations.

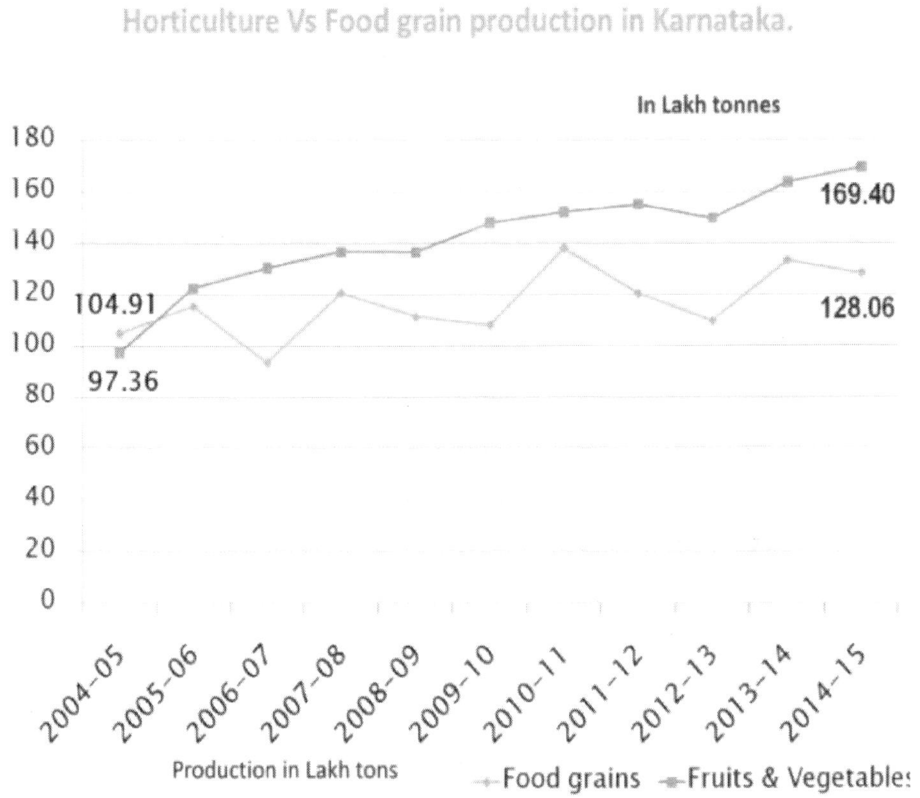

Strategy

To provide the required assistance and resources – policy action, inputs, technical knowledge, financial resources and infrastructure to strengthen FPOs. To provide over all support for the promotion of FPOs in sustainability by engaging qualified and experience resource institution (RI).

Concept

Focus on backward and forward linkages, through formation of farmer interest groups consisting of 20 farmers in each FIG. 50 such FIGs will form one FPO. Each FPO will concentrate on any one or a group of crops.

Organizing Farmers as FPOs

The collective bargaining power and wisdom of farmers needs to be improved. Department to encourage and facilitate farmers to organize themselves as FPOs through awareness meetings and discussion groups on advantages of FPOs. 92 FPOs were formed by end of the financial year 2015-16. Government to provide three year management support through specialized agencies with expenditure of Rs 30 lakh/FPO spread over three years.

Management support to include Provisioning of technical manpower @ 5 LRPs (Local Resource Persons), providing a paid CEO for three years. Adoption of each FPO by IIHR (Indian Institute of Horticulture Research) or UHSB (University of Horticulture Sciences Bagalkot) for technical domain support. Arranging visit of members of FPOs to other FPOs, Centre of Excellence in horticulture in the country and to other countries like Israel and China

State government would examine the possibility of one time grant of working capital out of state funds. Up to 10 lakh matching equity support from SFAC as per GOI guidelines on such support. State government would request SLBC to provide further working capital loan without collateral security. One time grant to be given from NABARD funds.

90 per cent subsidy by state government to horticulture FPOs limited to Rs 1 Crore per FPO for following infrastructural items: - Godowns, Collection centres, Integrated Pack-Houses, Processing units, Mechanizations, Custom Hire – Centres.

State government considering to make provision to provide certain percentage of scheme benefits to FPOs members. FPOs to themselves select beneficiaries for benefit of schemes in their areas. FPOs to work as oversight agency for such schemes - Certification of satisfactory work in its jurisdiction by FPO itself before payment can be made.

Empowering FPOs

FPOs to be empowered to play the role of input supplier for its members Grant of license by government for Fertilizer storage and sale to its members, Pesticide storage and sale to its members, Seed storage and sale to its members. They would also be empowered to have the much needed bargaining power in sale of their produce License as commission agent in APMC market, License as trader in APMC market, Provisioning of godown space on priority in APMC market.

Concept of FIG/FPO

To address the many challenging issues related to production, post harvest management and marketing, the department has identified a former member owned Horticulture Farmer Interested Groups. These groups were registered under Karnataka Society Registration Act 1960. The department of Horticulture is identified FIG/FPO as one of the most effective pathway to implement major schemes. *viz.,* CHD/VIUC/Bee keeping/Sujala and NHM *etc.* These schemes were implemented through cluster approach by establishing 80-100 ha large scale demonstration on single commodity with group action. Further, it is decided to converge these FIGs into FPOs at cluster/taluk level as commodity based FPOs such as Vegetables/ Fruits/Flowers based on the major crop grown in the taluk. The Department of Agriculture and Co-operation, GOI and SFAC had declared 2014 as " **Year of Farmer Producers Organizations (FPO)".**

Objectives

To promote economically viable self governing, democratic farmer producer organizations, to provide the required assistance and resources - policy action, inputs, technical knowledge, financial resources and infrastructure to strengthen FPO's, to provide over all support for the promotion of FPO's for sustainability by engaging qualified and experienced Resource Institution (RI) through SFAC, New Delhi. Formation of Farmer Producer Organization: Cluster identification → Baseline assessment and diagnostic study → Farmer Interest Group (15-20 farmer) → Min 50 FIG consisting 1000 Farmers → Farmer Producer Organization (FPO) → Executive Body (2 farmer/FIG) → General Body → General Manager → Staff → Local resource person (100 farmers).

Benefits to the FPO Members (Farmers)

Good quality seeds, fertilizers, pesticides and other inputs at low cost, Financial support for cultivation, purchase of machineries and other investments. The FPO will do the direct marketing which enable farmer to save times, transaction costs, weighment losses, price fluctuations and quality maintenance. FPO will promote good horticultural practices, proper post harvest management and marketing by providing technical training, demonstrations and exposure visits. The farmers are provided with channel of information related to marketing prices, consumer demand and subsidy programmes of the Government. The FPO will provide crop insurances to the farmers. By using FPOs as implementing agencies for various horticultural development programmes, especially RKVY, NFSM, ATMA, HOPCOMS *etc.* Extending the benefits of Central and State funded programmes in horticulture to members of FPOs on a preferential basis. By appointing FPOs as procurement agents for MSP operations for various horticultural crops.

FPO Service Model.

ASSUMPTION

No. of Farmers	1000
No. of villages	8-9
No. of Farmer Interest Groups	50
No. of members per FIG	20
Years of intervention	3
No. of FPO	1
Objectives	a) Organize small farmers into FIG and FPO b) Horticultural technology promotion c) Market linkage
Key Strategies	a) Maximum use of local resource personnel b) Preferably selecting areas where land and water related investment (watershed) is done in the past/on-going for value addition c) Leveraging from other resources

Mode of Implementation

Comprehensive Horticulture Development Programme (CHD) has been launched by Department of Horticulture from 2012- 13, through cluster approach involving 3-5 villages. Subject matter specialist and Farmer facilitators were appointed on contract basis to help the FIG members with forward and backward linkages, Large scale precision farming demonstration organized through FIG's in 80-100 Acres involving 20-25 farmers, Rs. 2.00 lakhs revolving fund were provided for multipurpose use including marketing. Similar kind of FIG's were also formed in the other schemes of the Department *viz.*, Sujala/VIUC/Bee keeping. Totally 1080 FIG's were formed with 36,634 members across the State. To strengthen these FIG's, through FPO's for further integration and consolidation for forward and backward linkages.

National Policy for the Promotion of Farmer Producer Organisations

VISION: To build a prosperous and sustainable Horticulture sector by promoting and supporting member-owned producer Organizations. Enable farmers to enhance productivity through efficient, cost-effective and sustainable resource use and realize higher returns for their produce. The collective action supported by the government, and fruitful collaboration with academia, research centers, civil society and the private sector.

Role of Central Government Institutions in Supporting FPOs

Department of Agriculture and Cooperation (DAC), Ministry of Agriculture, Govt. of India will act as the nodal agency for the development and growth of FPOs.

Small Farmers' Agribusiness Consortium (SFAC), a Society under DAC, will be the designated agency of DAC to act as a single-window for technical support, training needs, research and knowledge management and to create linkages to investments, technology and markets.

SFAC will provide all- round support to State Governments, FPOs and other entities engaged in promotion and development of FPOs. In particular, SFAC will create sustainable linkages between FPOs and inputs suppliers, technology providers, extension and research agencies and marketing and processing players, both in the public and private sectors.

The mandate of National Cooperative Development Corporation (NCDC) will be expanded to include FPOs in the list of eligible institutions, which receive support under the various programmes of the Corporation.

NAFED will take steps to include FPOs in the list of eligible institutions, which act on its behalf to undertake price support purchase operations.

DAC will work with Food Corporation of India (FCI) and State Governments to encourage them to include FPOs as procurement agencies under the Minimum Support Price (MSP) procurement operations for various crops.

DAC and its designated agencies will work with NABARD and other financial institutions to direct short and medium term credit for working capital and infrastructure investment needs of FPOs.

DAC will also work with all relevant stakeholders to achieve 100 per cent financial inclusion for members of FPOs and link them to Kisan Credit Cards.

DAC will work with Ministry of Corporate Affairs and other stakeholders to further clarify and strengthen provisions of the law relating to the registration, management and regulation of FPOs with a view to fostering fast paced growth of FPOs.

Role of State Government Institutions in Supporting FPOs

By declaring FPOs at par with cooperatives registered under the relevant State legislation and self-help groups/federations for all benefits and facilities that are extended to member-owned institutions from time to time.

By making provisions for easy issue of licenses to FPOs to trade in inputs (seed, fertilizer, farm machinery, pesticides *etc.*) for use of their members as well as routing the supply of agricultural inputs through FPOs at par with cooperatives.

By using FPOs as producers of certified seed, saplings and other planting material and extending production and marketing subsidies on par with cooperatives.

By suitable amendments in the APMC Act to allow direct sale of farm produce by FPOs at the farm gate, through FPO owned procurement and marketing centres and for facilitating contract farming arrangements between FPOs and bulk buyers.

By appointing FPOs as procurement agents for MSP operations for various crops.

By using FPOs as implementing agencies for various agricultural development programmes, especially RKVY, NFSM, ATMA *etc.* and extending the benefits of Central and State funded programmes in agriculture to members of FPOs on a preferential basis.

By linking FPOs to financial institutions like cooperative banks, State Financial Corporations *etc.* for working capital, storage and processing infrastructure and other investments.

By promulgating state level policies to support and strengthen FPOs to make them vibrant, sustainable and self-governing bodies.

Principles for Sustaining Farmer Producer Organisation Development

- Voluntary and Open Membership,
- Democratic Farmer Member Control,
- Farmer-Member Economic Participation,
- Autonomy and Independence, Education,
- Training and Information,
- Co-operation among FPOs,
- Concern for the Community

Impact Analysis of Production and Productivity of Commodities

Table 3.1: Karnataka State Horticulture Promoted FPOs Business Transaction

Sl. No.	District	Taluk	FPO Name	Output Business		
				Commodity Traded (Crop)	Quantity Traded MT	Total Output Business (Rs.)
1	Chikballapura	Sidhalghatta	Sadalamma HFPC Ltd	Red Rose Onion	312.76	7,976,740.00
2	Bagalkote	Hungund	Hungund	Pulses	861.00	6,008,962.00
3	Bangalore (R)	Rajghatta	Rajagatta HFPC Ltd	Vegetables	148.42	4,253,255.50
4	Vijayapur	Basawan-bagewadi	Basawan-bagewadi	Onion	418.00	3,970,000.00
5	Kolar	Mulbagal	Sri vinayaka HFPC Ltd.	Mango and Potato	151.00	3,709,802.00
6	Ramanagar	Magadi	Srigiripura HFPC Ltd.	Coconut and vegetables	149.25	3,239,161.00
7	Bellary	Hospet	Thungabadra HFPC	Dry Chilly	48.45	3,200,000.00
8	Ramanagar	Channapatna	Gombeyanadu HFPC Ltd.	Baby Corn.	368.46	2,602,358.40
9	Bangalore (R)	Tubugere	Tubugere HFPC Ltd	Vegetables	189.90	2,335,127.00
10	Haveri	Haveri	Bhoothayi HFPC	Mango	120.15	2,000,000.00
11	Bangalore (U)	Anekal	Anekal HFPC Ltd.	Capsicum, Carrot, Onion, Corindar	60.00	1,322,331.00
12	Davanagere	Davanagere	Devanagari HFPC	Banana, Pomegranate, Sambar cucumber	29.20	1,292,000.00

Sl. No.	District	Taluk	FPO Name	Output Business		
				Commodity Traded (Crop)	Quantity Traded MT	Total Output Business (Rs.)
13	Chikballapura	Chintamani	Sri Amara-narayana HFPC Ltd	Vegetables	47.28	1,282,607.00
14	Kolar	Srinivaspura	Shree Kodanda-ramaswamy	Mango, Tomato	130.87	1,097,690.00
15	Gadag	Gadag	Puttaraja-gavayi HFPC	Dry Chilly	0.00	725,956.76
16	Hassan	Belur	Yagachi HFPC	Vegetables	46.07	593,200.00
17	Tumkur	Tumkur	Hebburu	Banana	18.53	514,989.00
18	D.K.	Belthangadi	Navachethana HFPC Ltd.	Banana,Cocoa	8.97	473,149.00
19	Haveri	Hirekrur	Kumadwati	Onion and Maize	39.50	468,000.00
20	Belgaum	Belagavi	Siddeshwar HFPCL	Vegetables and Paddy	50.00	371,000.00
21	Tumkur	Sira	Gadinadu HFPC Ltd	Pomogranate	16.00	350,600.00
22	Gulbarga	Kamalapur	Grameena	Banana and papaya	54.00	344,330.00
23	Dharwad	Hubli	Uluvayogi HFPC	Mango	6.00	300,000.00
24	Chamaraj-nagar	Kollegala	Kollegala	0.00	0.22	294,848.00
25	D.K.	Bantwal	Pingara HFPC Ltd.	Banana	31.28	285,573.00
26	Chikka-ballapur	Bagepalli	Cheluru	Carrot	18.00	252,000.00
27	Dharwad	Dharwad	Kayakayogi HFPC	Mango, Vegetables	38.31	250,000.00
28	Bijapur	Indi	Shanteshwara HFPCL	Lemon	3.00	228,838.00
29	Davanagere	Channagiri	Shanthisagar HFPC	Banana	6.54	216,000.00
30	Chitradurga	Chitradurga	Sri Sirigere HFPC Ltd.	Maize	30.70	212,000.00
31	Tumkur	Pavagada	Pavagada HFPC Ltd	Pomegranate and Papaya	0.00	205,660.00
32	Shimoga	Sagara	Malenadu HFPC Ltd.	Pineapple	9.75	200,106.00
33	Belgaum	Gokak	Vivekanand	Vegetables	8.80	183,600.00

Sl. No.	District	Taluk	FPO Name	Output Business		
				Commodity Traded (Crop)	Quantity Traded MT	Total Output Business (Rs.)
34	Udupi	Karkala	Karkala HFPC Ltd.	Organic vegetables and oil seeds	3.17	163,775.00
35	Mysore	Nanjangud	Nanjangud HFPC Ltd.	Vegetables	17.00	148,898.00
36	Chitradurga	Holalkere	Hombelaku	Arecanut and maize	0.00	138,500.00
37	Bagalkote	Bagalkote	Kaladgi HFPCL	Sapota and Pomegranate	0.00	123,135.00
38	Yadgiri	Yadgiri	Dayasagar HFPC	Watermelon	0.00	108,500.00
39	Bidar	Humnabad	Kayakalpa HFPC	Ginger,Mango	12.86	107,108.00
40	Bagalkote	Jamkhandi	Jamkhandi HFPCL	Maize, Grapes	12.00	90,000.00
41	Chickmagalur	Ajjampura	Sri Guru Sidhara-meshwara HFPCL	Onion	10.20	86,929.00
42	Madikeri	Madikeri	Bhagandesh-wara HFPC	Coffee powder, Pepper, Honey, Cardamom, Coconut, Coconut Oil, Kachampuli Acrecanut Plates	3.00	60,460.00
43	Koppal	Koppal	Abhinava-shree HFPC	Mango	1.00	55,520.00
44	Tumkur	Gubbi	Hemavathi HFPC Ltd	Banana+ Coconut	2.00	46,227.05
45	Belgaum	Chikkodi	Nippani HFPCL	Baby Corn. GreenChilli	21.00	46,000.00
46	Chitradurga	Hiriyur	Vanivilasa	Aonla	1.00	36,200.00
47	Mandya	Malavalli	Malavalli HFPC Ltd.	Vegetables	1.50	25,360.00
48	Chamaraja-nagar	Chamaraja-nagara	Udigala HFPC Ltd.	Vegetables	0.00	15,552.00
	GRAND TOTAL				**5750**	**52,014,047.71**

Table 3.2: Business Details of Phase 1 Farmer Producer Organization (FPO) Promoted by Department of Horticulture, Govt. of Karnataka, India (*Rs. in Lakh*)

Sl. No.	*Districts*	*Input Sales **			*Output Sales***		
		No. of FPOs	*Average Transaction*	*Average Profit*	*No. of FPOs*	*Average Transaction*	*Average Profit*
1	Bagalkote	2	106.30	7.42	2	2.16	0.06
2	Bangalore (Rural)	3	74.50	4.37	1	1.20	0.03
3	Bangalore (Urban)	1	373.95	5.80			
4	Belgaum	3	69.87	3.74	3	5.71	0.55
5	Bellary	1	6.54	1.50	1	3.44	0.68
6	Bidar	1	3.48	1.18			
7	Bijapur	2	35.48	5.40	1	0.59	0.16
8	Chamarajanagar	2	10.36	1.26	1	0.60	0.02
9	Chickmagalur	2	117.58	4.72	1	2.50	0.07
10	Chikballapura	2	37.84	2.64	2	20.58	0.59
11	Chitradurga	2	55.48	3.09	2	3.45	0.09
12	Dakshina Kannada	2	40.36	5.34	2	6.22	0.29
13	Davanagere	3	80.40	6.36	3	2.95	0.15
14	Dharwad	2	29.64	2.54	2	3.32	0.08
15	Gadag	1	10.47	2.08			
16	Gulbarga	1	46.50	7.06			
17	Hassan	2	43.07	4.84			
18	Haveri	2	34.75	4.74	1	0.25	0.05
19	Kolar	3	38.45	3.22	3	9.05	0.71
20	Koppal	2	42.81	4.36	1	0.27	0.08
21	Madikeri	1	75.36	7.12	1	1.11	0.19
22	Mandya	2	25.16	2.09			
23	Mysore	3	36.91	1.66			
24	Raichur	1	69.65	9.41			
25	Ramanagar	3	114.52	1.60	1	0.85	0.05
26	Shivamoga	2	231.88	14.66			
27	Tumkur	3	57.17	4.06	2	4.91	0.23
28	Udupi	1	45.00	3.00			
29	Uttara Kannada	1	65.22	2.40			
30	Yadgiri	1	5.47	0.45			

* Inputs sales include fertilizers, plant protection chemicals, seeds, tarpaline, CHC, cattlefeed, mulching sheet, drip material *etc.*

** Output sales include trading of processed and unprocessed Horticulture produce.

Table 3.3: Business Details of Phase 1 Farmer Producer Organization (FPO) Promoted by Department of Horticulture, Govt. of Karnataka, India (Rs. in lakh)

Sl.No.	*Districts*	*Input Sales **		*Output Sales***	
		No. of FPOs	*Average Profit*	*No. of FPOs*	*Average Profit*
1	Bagalkote	2	7.42	2	0.06
2	Bangalore (Rural)	3	4.37	1	0.03
3	Bangalore (Urban)	1	5.80		
4	Belgaum	3	3.74	3	0.55
5	Bellary	1	1.50	1	0.68
6	Bidar	1	1.18		
7	Bijapur	2	5.40	1	0.16
8	Chamarajanagar	2	1.26	1	0.019
9	Chickmagalur	2	4.72	1	0.07
10	Chikballapura	2	2.64	2	0.59
11	Chitradurga	2	3.09	2	0.092
12	Dakshina Kannada	2	5.34	2	0.29
13	Davanagere	3	6.36	3	0.15
14	Dharwad	2	2.54	2	0.08
15	Gadag	1	2.08		
16	Gulbarga	1	7.06		
17	Hassan	2	4.84		
18	Haveri	2	4.74	1	0.05
19	Kolar	3	3.22	3	0.71
20	Koppal	2	4.36	1	0.08
21	Madikeri	1	7.12	1	0.19
22	Mandya	2	2.09		
23	Mysore	3	1.66		
24	Raichur	1	9.41		
25	Ramanagar	3	1.60	1	0.05
26	Shivamoga	2	14.66		
27	Tumkur	3	4.06	2	0.23
28	Udupi	1	3.00		
29	Uttara Kannada	1	2.40		
30	Yadgiri	1	0.45		

* Inputs sales include fertilizers, plant protection chemicals, seeds, tarpaline, CHC, cattlefeed, mulching sheet, drip material *etc.*

** Output sales include trading of processed and unprocessed Horticulture produce.

Table 3.4: State Govt Looking Forward to Receive Value Chain Projects in following Fruits, Vegetables and Flowers

Crop Name	*Area Covered (in Lakh Hectares)*	*Crop Name*	*Area Covered (in Lakh Hectares)*
Mango	1.73	Tomato	0.62
Banana	1.02	Green Chillies	0.43
Pomegranate	0.18	Flowers	0.30
Grapes	0.20	Spices	2.00
Potato	0.42	Cashew nut	0.71
Onion	1.60	Coconut	4.95

State govt would seek support of CII, FKCCI, World Economic Forum and other organisations to facilitate such partnership. Individual companies generally working in this area are being requested through formal letters to participate in this scheme.

Table 3.5: Income and Employment Generation of IFSD Model

Farming System	*Expenditure*	*Gross Income (Rs. ha^{-1})*	*Net Returns (Rs. ha^{-1})*	*Employment Man Days $Year^{-1}$*
Arable	14,171	38,264	24,093	257
MF with 2 cows	34,972	72,640	37,668	374
MF with 2 buffalo	47,257	71,545	24,288	390
MF with 2 cow + fish	35,170	76,064	40,894	374
MF with 2 buffaloes +fish	47,455	74,969	27,514	390
MF with 2 cows + 15 goats +10 poultry + 10 ducks + fish	43,311	88,222	44,911	380
MF with 2 buffaloes +15 goats + 10 poultry + 10 ducks + fish	55,596	87,127	31,531	396

Dairy (Breed: HF cow)

Poultry birds (Breed: Swarnadhara)

Installation of Bee Hive Box

Vermicompost Unit

Enterprise-wise Contribution to Total Net Income (Rs).

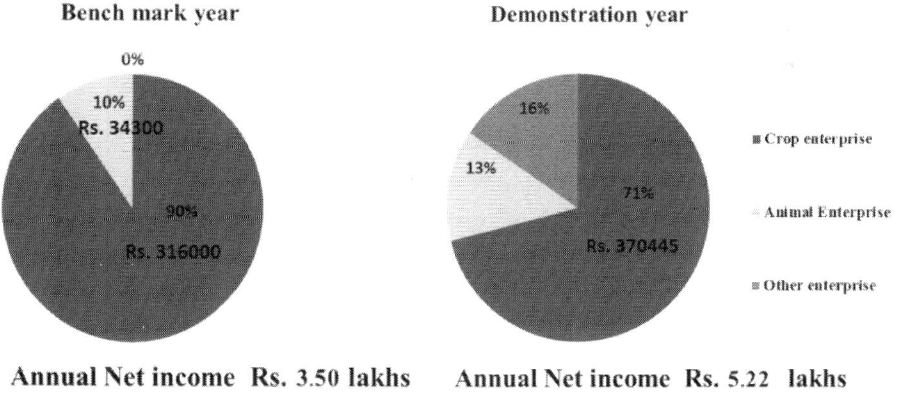

Increase in Annual Net Income over Bench Mark Year: Rs. 1.65 lakhs

Table 3.6(a-c): Economic Analysis of the Major Horticulture Crops of Large Scale ICM Demonstration Established through CIG's/FIG's

(a) Economics of Banana Production under Precision and Non Precision Farming

Sl. No.	*Particulars*	*Non Precision Farming*	*Precision Farming*	*Per cent Increase Over Non Precision Farming*
1	No. of plants/Ha	3000 (1.8 m x 1.8 m)	3333 (2.0 m x 1.5 m)	11.10
2	Average weight of fruit bunch	25-30 kg	40-45 kg	50.00
3	Yield/Ha	90 t	149 t	65.55
4	Total Cost/Ha	Rs. 1,98,605/-	Rs. 3,46,927/-	74.70
5	Gross Returns/Ha	Rs. 10,80,000/-	Rs. 20,86,000/-	93.15
6	**Net Returns/Ha**	**Rs. 8,81,395/-**	**Rs. 17,39,073/-**	**97.30**

Success Story of FIG's/CIG's of Department of Horticulture through Large Scale Banana ICM Demonstration.

(b) Economics of Tomato Production under Precision and Non Precision Farming in Open Condition

Sl. No.	*Particulars*	*Non Precision Farming*	*Precision Farming*	*Per cent Increase Over Non Precision Farming*
1	Seed Rate/Ha	75-100 g	100-120 g	20.00
2	Yield/Ha	40-45 t	75-80 t	77.77
3	Total Cost/Ha	Rs. 45,950/-	Rs. 1,59,788/-	247.74
4	Gross Returns/Ha	Rs. 1,35,000/-	Rs. 3,60,000/-	166.66
5	**Net Returns/Ha**	**Rs. 89,050/-**	**Rs. 2,00,212/-**	**124.83**

Success Story of FIG's/CIG's of Department of Horticulture through Large Scale Tomato ICM Demonstration.

(c) Economics of Rose Production under Precision and Non Precision Farming in Open Condition

Sl. No.	*Particulars*	*Non Precision Farming*	*Precision Farming*	*Per cent Increase Over Non Precision Farming*
1	No. of plants/Ha	10000 (1.0 m x 1.0 m)	13,333(0.75 m x 0.75 m	33.33
2	No. of cut stems/ Plant/Year	30-40	50-60	57.14
3	Yield/Ha/Year	3,50,000cut stems	5,50,000cut stems	57.14
4	Total Cost/Ha	Rs. 1,32,048/-	Rs. 2,36,974/-	79.46
5	Gross Returns/Ha	Rs. 4,37,500/-	Rs.11,00,000/-	151.42
6	**Net Returns/Ha**	**Rs. 3,05,452/-**	**Rs. 8,63,026/-**	**182.54**

Success Story of FIG's/CIG's of Department of Horticulture through Large Scale Rose ICM Demonstration.

Minimum Support Price (MSP) Procurement Operations for Sunflower – Assisted by NAFED,Central Government

Some of the Issues Faced by the Horticulture Farmers

Farmers gets low share in the Crop Value chain., Intermediaries benefit most Perishable nature of Horticulture produce force farmers to sell them at throw away prices. Protest by farmers against collusion between traders bidding for produce in markets, Lack of proper infrastructure and mechanization tools increases the cost of production. Individual farmers lack bargaining power, Institutional buyers doesn't have mechanism to reach individual farmers and aggregate the produce.

Govt. has taken Measures to Handle above Problem

Reforms in working of APMC market, Liberalization of APMC Act to benefit farmers. Horticulture Dept. organize farmers as Farmer Producers Organizations (FPO) and provide them with – Input support at subsidized prices, Infrastructure support, Management Support. Connecting farmers to Corporate Sector to get them proper value for their produce.

A Revolution in Making – Public Private Partnership for Integrated Horticulture Development (PPP-IHD)

Public Private Partnership for Integrated Horticulture Development As per GOI framework of PPP-IAD

Tripartite partnership- Farmers(FPO), Corporate and state government Collaboration from seed to produce including Supplying of Seeds, fertilizers and pesticides, Mechanization tools, Extension and training, Buyback of produce at direct purchase centres with immediate RTGS payment, but no compulsion on the part of the farmer to sell his produce to the company.

PPP - IHD Features

Companies to submit Detailed Project Reports (DPR) covering all aspects of a Crop Value Chain - from inputs to purchase of produce. Overall project cost sharing is expected to be 1:1:1 between company, government and farmers. Govt share for any component of DPR to be as per the guidelines of NHM, NMSA *etc.* Ex providing drip irrigation component for grapes project would have govt support to the extent of 90 per cent. Whereas a component on provisioning of packhouse may attract govt support to the extent of 50 per cent support under NHM.

PPP IHD – Path and Goal

A state level committee chaired by Principal Secretary Horticulture will examine and approve the DPRs. Initial funding of Rs 10 Crore provided for the current year under RKVY. It is expected that the benefits of project would be as follows Increase in productivity by 35 per cent, Increase in purchase price per unit by 10 per cent, Increase in overall income by 10 to 30 per cent.

Coorg Coffee (100& 200 gm) Coorg Cardamom (50&100gm) Coorg Vinegar/Kachampuli (180&750ml)

Conclusion

The registered FPOs have been empowered by providing input license for sale of Fertilizers, Seeds and pesticides to the members for reasonable rates and also provided marketing license to sale the produce in all the APMC centres of the state and also outside the state and country with company registered license which will empower the members of FPOs to avoid middlemen at all levels and encash higher net profit with lesser cost of inputs, further the Department of Horticulture is also empowering the members of FPOs by providing 90 per cent subsidy for establishing infrastructure facility and custom hiring centre. Impact of technology transfer and adoption of technology has shown highest income per unit area has indicated in tables through precision farming in comparison with non-precision farming. The percentage increase in net income per unit area has clearly indicated the potentiality of income generation empower the farmers.

As indicated in the Tables, it is evident that the net profit earned by each FPO is more than crores from sale of inputs (seeds, fertilisers and pesticides *etc.*) and net income is also earned from customised center of equipment. Further, the net profit was also earned from the sale of the produce from members. This shows the total empowerment of the FPOs for income generation and employment opportunities at rural setup.

Among these 90 FPOs 9 FPOs linked to corporates under PPPIHD projects, where companies will support the farmers right from seed to marketing. The established FPOs have completed 2-3 years. The performance of FPOs has creating very good economic, social and group dynamic qualities among members which helps in empowerment of FPOs which indicates the total turnover will be around one crore per FPO. Further there is good scope for processing and value addition for their produce.

The PPP-IHD projects has also shown good results by linking corporates for procurement of the high quality produce through FPOs which was produced by the members of FPO. This has also been highlighted through establishing small scale processing units, which has clearly shown the scope for entrepreneurship among FPOs.

As indicated in the Farming System the additional income generation and employment generation is quite evident from IFS system from different source such as dairying, poultry, piggery, beekeeping, sericulture *etc.* can also be linked to FPO system for sale of produce and also value addition. This system is well established to accommodate all kinds of enterprise which can be encouraged for rural setup through Farmers Producers Organizations. This system can be strengthened with effective forward and backward linkages. This can enhance the value chain and marketing linkage, which needs to develop different extension strategies and policy imperatives in respect to IFS system.

Chapter 4

Strategies for Efficiency Enhancement of Small Farmer Production Systems

S. Prabhu Kumar[1], *V. Veerabhadraiah*[2] *and G. Eswarappa*[3]

[1]*Former Zonal Project Director, ICAR- Zonal Project Directorate, Zone-I, Ludhiana and Zone VIII Bengaluru*

[2]*Former Director of Extension, UAS, Bengaluru and President, Institution of Agricultural Technologists, Bengaluru*

[3]*Former Director of Extension, UAS, Bengaluru and Vice President – ISEE (South Zone) and Vice President, Institution of Agricultural Technologists, Bengaluru*

Confession of a Small Farmer

> *"I have a policy that I should get revenue from my farm daily, weekly, monthly, quarterly, half yearly and yearly. I follow this practice firmly therefore I plan the crop cultivation accordingly. I have a complete integrated farm. I grow flowers, vegetables, fruits, exotic vegetables in poly house, dairy, pig, poultry and fish farming".*

Small-scale agriculture has changed dramatically since the time of the Green Revolution. Despite predictions by many economists suggesting increased land consolidation, the reverse has held true in many places. Shrinking small-scale farm sizes have led the rural poor to diversify in to off-farm income sources, further compromising agricultural output. Added to these trends emerging stressors like climate change and growing populations and you have the makings of a perfect storm facing small-scale producers. Recognizing that not all small-scale farmers are the same and adopting targeted measures to support them is an important step in adapting to this new era of smallholder farming.

Small-scale farms are still highly efficient (*i.e.* the smaller the land size the greater the productivity) although *labour* productivity remains low. A proliferation of agricultural cooperatives and other collective-action arrangements today has achieved some success in allowing small-scale producers to gain access to markets, credit and inputs. Supporting smallholder farmers remains a good strategy for poverty alleviation and food security - as they represent a social agenda in their own right - but unless small-scale farms succeed in generating linkages to off-farm sectors then their outlook in terms of economic growth remains poor.But not all small-scale farms are created equal.

Types of Small Farmers

- ☆ **Business oriented small farmers :** (*i.e. those that are already linked -or nearly linked - with markets and quality inputs*)
- ☆ **Smallholders in transition:** (*i.e. those part time farmers that are likely to leave agriculture*)
- ☆ **Subsistence oriented small farmers:** (*food insecure, marginalized or ethnically discriminated farmers*)

Support measures, that could be targeted according to these distinct typologies include for example,

- ☆ Improved technologies and market access for business oriented small farmers.
- ☆ Training and support for entrepreneurs and small businesses to support successful exits from agriculture for smallholders in transition.
- ☆ Safety nets and empowerment of women and vulnerable groups for subsistence-oriented small farmers.

Present Status and Needed Approach

Status of present farming include smaller acreage farms, compared to national average, increasing land prices per acre, increasing production costs, growing disconnect between value of farm production and retail food prices. Under this situation we should concentrate on the strategies for promoting **PRIDE, PROFIT and PRESTIGE (PPP)** among the farming community.

Indian agriculture is Complex, Diverse and Risk Prone. The agricultural development activities should address all these three issues for sustainable livelihood security for the farmers. Under this situation, we should concentrate especially on better prices, cost reduction, mitigating risk and improve supply chain efficiency. The focus should be on addressing 3Ms in agriculture *i.e.*, **MARKET, MONSOON and MINDSET.**

Strategies

The need of the hour is changing from supply driven extension to demand driven extension. Mainly the strategies for promoting PPP should be more on empowering pride of ownership with profit and prestige to the farming occupation. Under this situation, technocrats in agriculture should think beyond the technology delivery as farmers are very much conversant with technology part and they need more empowerment in taking the farming as a way of business rather than a way of life. In this regard, the following strategies namely Speciality Agriculture, **Smart Agriculture for Smart Farmers and Precision Agriculture.**

Specialty Agriculture

In agriculture, development can happen in three phases. They are:

I. Primary Agriculture
II. Secondary Agriculture
III. Specialty Agriculture

Primary Agriculture refers to production of agricultural commodities. Secondary Agriculture is the process of making money with the agricultural commodities through processing, value addition and marketing. Speciality Agriculture is adding value to a product or process.

Speciality Agriculture can be briefly described as non-traditional agribusiness products, including commodity items not traditional, value-added agribusiness products or processes and traditional agribusiness products which are marketed or distributed using new/innovative channels.

Examples of Specialty Agriculture include extended season produce, identity preserved grains, organic products, bio-fuels, further processed food items, alternative forest products and agritourism/agritainment.

Importance of Specialty Agriculture covers potential for greater return on investment or greater return per acre than traditional commodity items, allows agribusiness producers to take advantage of competitive advantages gained through diversification and/or differentiation and greater long-term growth potential for specialty agriculture products than traditional commodity items.

Specialty Agriculture trends around creating value vs. capturing value. Capturing value is a function of altering the distribution of value in the agricultural production chain, or by cutting costs(*vertical integration, FPOs*). Creating value is a function of altering the actual or perceived value to a customer (*unique or branded products, identity preserved or specialty products, bundling such as agri-tourism*).

In other words Speciality Agriculture confines to precision agriculture (*protected environment, assured production, hi-quality products, assured market, higher rates*) as compared to traditional *farming* (*depends on nature, uncertainty of crop, low quality produce, non remunerative prices, lack of markets*).

Vision of Speciality Agriculture is to ensure at least a minimum income of Rupees One thousand per day by a family from one acre of land. Farmers can achieve this if they are supported with relevant advice on production and marketing. Farming is a lucrative option and in speciality agriculture educated youth should come forward to farming. Small farmers, who dominate Indian agriculture, need to be organized and the organizational structure of a producer company will help small farmers to realize better prices. Through adoption of high-tech farming, one can manage land, labour, power and other inputs efficiently.

Smart Agriculture for Smart Farmers

Climate Smart Agriculture is not a new concept but has been evolved as a result of complex problems arising due to climate change which can be mitigated by efforts, analysis of situation (diagnosis based prescription) and programmes that meet such requirement and flexible enough to change as per required situation. The focus has to be small stakeholders and uneducated farmers, women farmers and rural youth.

Approach: It needs changes in behaviour and strategies as well as changes in the usual timing of agri practices and adopting new crops/switch crops depending upon changes in weather pattern during season.

Needed Support

- Sharing technical knowledge.
- Providing credit, subsidy and safety net.
- Monitoring and co-ordination of different agencies
- Providing information to farmers
- Available option for crops based on soil health and moisture analysis, for markets – local and district levels.
- About climate – weather forecasting - long – medium – short term day to day. This is should be supported by agro - advisory
  a) based on forecast
  b) after unforeseen climate event has taken place like heavy rains and flooding of fields.
- About opportunity to have multiple source of income.
- Different players in agriculture sector need completely different types of climate advise.
- The new extension mechanism has to keep the following in view.
  a) First is need prior to commencement crop season – what they should grow – pre kharif (monsoon) and pre-rabi (winter) based on prevailing climate parameters.
  b) This has to be followed by intra – weather agro advisory – based on unforeseen changes in weather during season (contingency plan).
  c) Finally at post harvest stage – where they sell products or store it till the market price is improved.
  d) Farmer also needs to know what is market demand - both as regards to quality and in terms over all requirement and possible market price. This will enable him to grow right type of crop.

e) Similar advisory is needed for livestock management and fisheries. Hence, entire exercise is multi- dimensional and extension officer have to be trained for that.

Key Recommendations for Smart Agriculture

- ☆ Unpredictable and erratic climatic patterns resulting from climate change will affect crop production. This will have an impact on farmer livelihoods and food availability. Climate-Smart agriculture provides management options to farmers to both adapt to, and mitigate climate change and maintain his income and look for opportunity to increase it.
- ☆ Crop production must adapt - crop varietal selection, plant breeding, cropping patterns and ecosystem management approaches and become resilient to changes (frequency and intensity).
- ☆ Crop production can contribute to mitigating Climate Change by reducing green house gas (GHG) emissions - for example by reducing the use of/judiciously using inorganic fertilizers, avoiding soil compaction or flooding to reduce methane emissions (*e.g.* in paddy rice systems) and sequestering carbon (*e.g.* planting perennial crops and grass species).
- ☆ Farmers are the primary custodians of knowledge about their environment, agro-ecosystems, crops and cropping patterns, and local climatic patterns. Adapting cropping practices and approaches will be related to local farmers' knowledge, requirements and priorities. Sustainable crop production provides farmers with options for farming sustainably, taking into account the local ecosystem and how to follow selection of crops which can be sustained by soil – based on soil health and moisture analysis of their land.
- ☆ Integrated approaches — such as crop-livestock systems, rice-fish systems and agroforestry — diversify food sources and consequently strengthen the resilience of farmers' livelihoods. They also provide opportunities for mitigating climate change but more precisely also increase their income level and in worst circumstances maintain it.
- ☆ Establishment of Bio Pharmacy in order to provide quality bio products like *trichoderma, pseudomonas*, vermicompost, botanicals and microorganism, *etc.*, for promoting organic agriculture.

Precision Agriculture

Precision Agriculture or Precision Farming (PF) is a concept of using the new technologies and collected field information, doing the right thing, in the right place, at the right time. Collected information may be used to more precisely evaluate optimum sowing density, estimate fertilizers and other inputs needs, and to more accurately predict crop yields. It helps in avoiding unwanted practices to a crop, regardless of local soil/climate conditions, *i.e.*, it reduces labour, water, inputs such as fertilizers, pesticides *etc.* and assures quality produce.

Role of Farmer Associations

The farmers associations appear to perform two vital roles in the dissemination and success of PF technology. The associations serve as nodes for exchanging knowledge and information. They also help farmers obtain better value for produce as well as inputs. The associations perform a vital support function as information nodes. The association holds regular monthly meetings to discuss marketing and other issues. The associations act as demonstration vehicles to disseminate both knowledge and information about the techniques and the impact that they have on cultivation and post-harvest results. They appear to help reduce the knowledge and information deficit. The associations also seem to help the farmers obtain better value by improving their negotiating position vis-à-vis buyers or input providers. The organized markets increasingly prefer to deal with farmer associations as it helps to eliminate risks of delivery failure while providing a greater assurance of quality. This is also beneficial to the farmers as it helps them to secure better value by costing out delivery failures and in-transit damage to produce out of the revenue. By assuring minimum quality through proper grading and sorting, associations help farmers obtain better average prices than comparable produce sold without the association's involvement. The associations also help the farmers to negotiate better price for inputs such as fertilizers, pesticides, seeds, *etc.* by guaranteeing minimum quantity, as well as negotiating for or arranging *timely* supply of inputs. The role of the association was vital in facilitating in the marketing of the produce. The association also acted as the distributor or dealer for several agri-product corporations and sold input materials such as seeds, fertilizers, plant protection materials, and other agriculture inputs. The products sold were of better quality and at cheaper rates.

Beyond Association Farmers

The impact of the PF must be noticeable in terms of the extent to which the PF technologies travelled beyond the beneficiary farmers. The manner in which the precision farming technologies travelled to the farming community and the impact they had on the farmers. In terms of the travel of technologies, travelled unmodified from the producing domain to the receiving domain (beneficiary farmers). The extent of travel was complete in the sense that all beneficiary farmers adopted and continue to use this technology.

Under precision farming adding value to social capital is the highlight. We have to allow the farmers to understand the ground realities and involve them in all the decision making process. To achieve this, we have to adopt farmer-to-farmer strategy which will result in bargaining power, negotiating power, commercial activity, (total quality management, market study, organised visits, capacity development, self sustainability and (develop linkages.

Hence, the output of precision agriculture will be

- ✰ Doubling the productivity of crops
- ✰ Around 95 percent marketable produce with less unmarketable produces

- ☆ High end quality parameters and 30 per cent premium price in the market
- ☆ Water economy 30 to 40 per cent
- ☆ Electricity economy 50 per cent
- ☆ Extended period of harvest to match lean season in the market
- ☆ Less labor dependence
- ☆ Extended shelf life
- ☆ Roughly 25 per cent more weight per unit volume for the produce
- ☆ Empowerment of farmers - (*Technical, Economical and Social empowerment*)

Institutional Approach to Sustain the Technology Management among Small Farmers

Approach 1-Community Managed Resource Centre (CMRC) (Knowledge Management System Promoted by KVK)

CMRC is a non-profitable local level institutions managed by the community based organizations for sustainable development. It provides INFORMATION, TECHNOLOGY and SERVICES to Member CBOs and Farmers with the support of the ICAR- KVK Erode, Tamil Nadu.

The Centre is ably backed by trained/Experienced with Village Level Technocrats (VLTs) in various fields – Agriculture, Animal Husbandry and Institution Building

Structure of CMRC

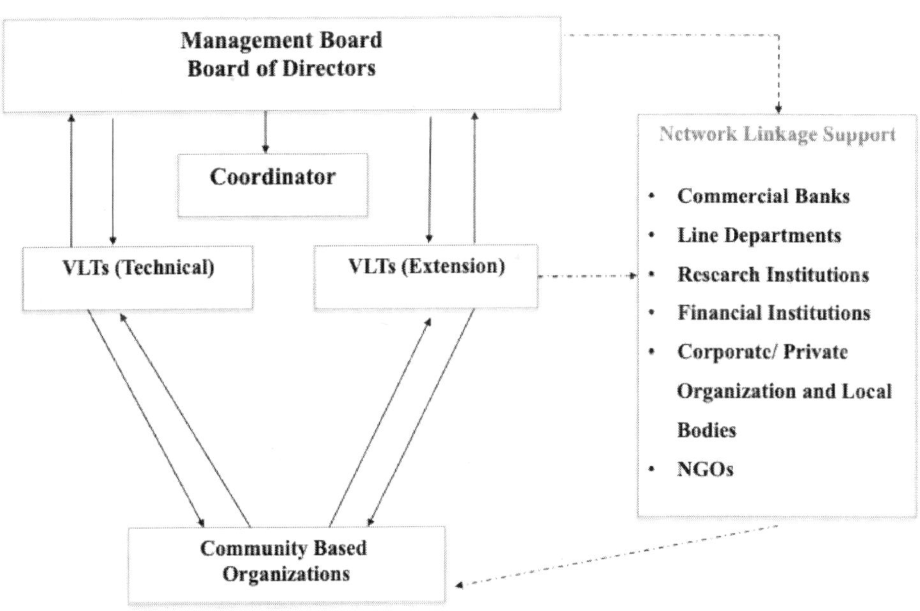

Membership

- Restricted to primary groups (*i.e.* membership groups and not representative/apex groups)
- Restricted to groups that are graded as 'well-functioning'
- Membership is secured through initial subscription and annual renewal charges
- Recommended membership is generally between 75 – 125 CBOs

Features of CMRC

- Vision/Mission
- OrganizationalManagement
- Financial Management
- OrganizationalAccountability
- Linkage
- Learning and Evaluation

Objectives

- To cater to the information needs of local communities on a regular basis.
- To ensure that local communities have access to need-based services on a regular/payment basis.
- To promote local community development in as many sectors and as many creative ways as ethically and economically feasible.
- To support its members (community based organisations) on a regular basis for :
  - Group grading
  - Maintenance of books and records
  - Training
  - Audits
  - Linkages with resource institutions
  - Any other routine/new services

Details of CMRCs ubder KVK Erode

- No. of CMRCs: 7

  (*Bargur, Anthiyur, Kallipatti, Talawadi, Arappalayam, Germalam and Kadambur*)
- No. of Members: 5,904
- No. of CRPs: 71

Modalities of Functioning

1. *Development of Village Level Technocrats (VLT) (Knowledge Management System)*

- ☆ Community Resource Persons: The CMRC identified Village Level Resource person to train them on Institution Building and its Sustainability aspects
- ☆ *Krishi Mitra* : Village Youths identified and trained on Farm Technologies and its management aspects
- ☆ Animal Health Promoter: Rural Youths identified and trained through professional institution on Animal Health and Welfare

Village Level Technocrats (VLTs) in the DistrictThe Village Level Technocrats attached with Community Managed Resource Center (CMRC) in 3 Taluks

Sl.No.	Name of Taluk	Name of VLTs	No. of VLT
1	Sathyamangalam	Community Resource Person	11
		Krishi Mitra	2
		Para Veterinary Professionals	13
2	Gobichettipalayam	Community Resource Person	3
		Krishi Mitra	6
		Para Veterinary Professionals	9
3	Anthiyur	Community Resource Person	6
		Krishi Mitra	5
		Para Veterinary Professionals	11
4	Perundurai	Krishi Mitra	3
		Para Veterinary Professionals	2
Total VLTs			71

2. *CMRC'S Services*

Technology Management Services

- ☆ Ensures quality seeds and saplings in right time
- ☆ Market-led information
- ☆ Kisan Mobile Advisory Service through KVK
- ☆ Production of on-farm inputs at site
- ☆ Skill training on processing and value addition of agricultural produces
- ☆ Selection of appropriate beneficiaries for implementing programmes
- ☆ Capacity building on latest technologies for improving productivity
- ☆ Linkage with various research institutes and line departments
- ☆ Formation of farmers/commodity based groups
- ☆ Documentation of Impact/Case studies

Animal Welfare

- ☆ Ensure the veterinary services through Community Resource Persons (Para-vet)
- ☆ Conserve the indigenous breeds
- ☆ Technologies provided to efficiently managed the cattle resources
- ☆ Production of concentrate feeds and mineral lick
- ☆ Organizing the camps on regular basis
- ☆ Access of veterinary services - Selection of Rural Youths for attaining par-vet training
- ☆ Upgrading the local breeds
- ☆ Encourage the farmers to produce fodder seeds on buy back

Farm Innovation

- ☆ Identification of local innovative farmers
- ☆ Innovations identified
  - ❒ Mechanism for managing cattle resources
  - ❒ Clod breaker
  - ❒ Castor Fermented Solution trap
  - ❒ Turmeric Harvester
  - ❒ Ground nut pod striper
  - ❒ Power weeder for drylands
  - ❒ Honey wax separator
  - ❒ Organic liquid manure preparation
  - ❒ Stone remover from field

General Services

- ☆ Creating awareness on legal rights
- ☆ Monitoring the members CBOs (Functioning and Grading)
- ☆ Organizing seasonal camps (Human and Livestock)
- ☆ Organizing trainings for landless labourers for improving their livelihood status
- ☆ Information related to on going programmes implemented by various line departments

Marketing Services

- ☆ Identification and promotion of Agri/Entrepreneurs
- ☆ Institutional linkages
- ☆ Train the farmer/Entrepreneurs on processing and value addition
- ☆ Formation of Agri preneurs and entrepreneurs groups

- ☆ Facilitation and formation of FPOs
- ☆ Encourage to participate in the exhibition organized at gross root level and state level
- ☆ Resource linkage: Technical/Raw material and Financial/Market information

Approach 2: PPP Mode for Fodder Promotion in Tamil Nadu – An Innovative Approach

Public Private Partnership in Fodder Seed Production started with signing of MoUby Farmers – KVK - Dept. of Animal Husbandry, Tamil Nadu. The role of partners include Farmers (Fodder Seeds Producers), Krishi Vigyan Kendra, Namakkal (Co-ordinator) and Dept. of Animal Husbandry, Tamil Nadu (Purchaser). The objective is to develop a standard for fodder seeds production for livestock and to increase the economic standard of the farmers through production and supply of fodder seeds.

Fodder Seeds Production Agreement Plan - Salient Features

- ☆ Farmers should produce the fodder seeds such as grasses, cereals, legumes and tree fodder as per specifications with 10 per cent moisture level and handover the same to the KVK.
- ☆ KVK should train and demonstrate on fodder seed production to the fodder seed producers and check the quality of seeds before supplying to Dept. of Animal Husbandry.
- ☆ Department should inform their needs to KVK in advance in order to procure seeds from the farmers and then supply for which Department should pay the amount within thirty days from the date of purchase.
- ☆ A 15 per cent service charge will be credited to KVK for conducting training, demonstration, seed testing, cleaning of seeds, packing and forwarding.

Coverage of Farmers Signed MOU for Fodder Seed/Setts Production in Tamil Nadu

Sl.No.	*District*	*No. of Farmers*
1	Namakkal	51
2	Salem	09
3	Trichy	06
4	Dharmapuri	05
5	Erode	04
6	Vellore	04
7	Dindigul	02
8	Krishnagiri	02
9	Cuddalore	02
10	Karur	01
11	Tiruppur	01

Sl.No.	District	No. of Farmers
12	Theni	01
13	Villupuram	01
14	Pudukottai	01
	TOTAL	**90**

Technologies

- ✰ Grasses - *Bajra Napier hybrid grass-CO (CN).4, Guinea grass-CO(GG).3*
- ✰ Cereals - *Fodder Sorghum – COFS 29, Fodder maize – African tall*
- ✰ Legumes - *Hedge lucerne, Fodder cowpea – COFC8*
- ✰ Tree fodders – *Subabul, Sesbania*

Technological Interventions

- ✰ Training
- ✰ Exposure visits
- ✰ Mechanization
- ✰ Farmer Field Schools
- ✰ Demonstrations

Production and Area coverage due to the seed production

Fodder Crop	Qty. Supplied	Area Covered (Acre)	Value (Rs. in lakh)
Bajra Napier Hybrid grass-CO (CN)4	12 lakh slips	82.00	3.79
Fodder Sorghum (Multicut)-COFS.29	19311 kg Seed	9656.00	77.24
Desmanthus	8910 kg Seed	1155.00	44.55
Subabul	717 kg Seed	187.00	1.81
Sesbania	561 kg Seed	239.00	3.26
Fodder cowpea	135 kg Seed	17.00	0.20
Anjan grass	99 kg Seed	10.00	0.35
Fodder maize	125 kg Seed	16.00	0.09
Stylo	141 kg Seed	14.00	0.49
Lucerne	24 kg Seed	5.00	0.13
Guinea grass	10000 Slips	1.00	0.10
TOTAL	**30 ton Seed** **12.10 lakh slips**	**11088.00**	**130.00**

Success of PPP in Fodder Promotion owes to Convergence

- ✰ TNAU Coimbatore,

- ☆ Dept of Animal Husbandry - Tamil Nadu.
- ☆ *Aavin* – Tamil Nadu.
- ☆ District Water Shed Development Agency.
- ☆ KVKs, VUTRCs, Research Farms,
- ☆ NDDB, Erode.
- ☆ *Pudhu Vazhvu* Scheme –Tamil Nadu.
- ☆ M.S.Swaminthan Research Foundation.
- ☆ Dept. of Forest –Tamil Nadu.

Achievements of the Approach

- ☆ Ninety farmers from 14 districts have signed MoU for fodder seed production.
- ☆ Thirty tons of fodder seeds worth Rs.1.3 crore covering 11088 acres of fodder area has been supplied.
- ☆ On an average farmer earn Rs.1.0 lakh/acre from fodder seed production excluding the fodder.
- ☆ High incidence of seed shattering in COFS -29 reported by KVK, Namakwa to TNAU breeders and a non-shattering type evolved for seed production.

Conclusion

Many interventions were made to make total farm production increased for small farmers, but still a well-equipped new approach has to be formulated.

- ☆ A systems approach is basic to understanding the concept of integrated crop livestock-fish-forestry development at small farm level.
- ☆ Small farmers are reluctant to take risks and are not enthusiastic about new practices. Therefore technology should not be expensive but should be generated based on the farming systems and its available indigenous resources. However, proper utilization of indigenous knowledge, methods, labour and material should be emphasized.
- ☆ Credit, marketing, extension service, technical backstopping and other important institutional supports should accompany the technical know-how.
- ☆ There is a need to develop coherent guidelines for developing various forms of crop-livestock-fish-forestry integration models. This is important as socio-economic conditions differ from one farm to another farm.
- ☆ A strategy of importance is to concentrate on development of on-farm production through proper types of livestock, crops including tree, vegetable and forage crops most suited to the agro-climatic and socio-economic conditions of the individual representative farms.

Chapter 5

Convergence of Farmer Led Innovations and Agripreneurship for Profitable Farming: Prospects and Experiences

Manjeet Singh Nain[1], Rashmi Singh[1], J.R. Mishra[1] and Anjani Kumar[2]

[1]Principal Scientist, Division of Agril Extension, ICAR-IARI, New Delhi – 110 012

[2]ICARAagricultural Technology Application Research Institute, Patna

Farmers are creative and generate relevant local innovations which comes from a variety of sources including *intuition, dreams, experience at work, training, the ideas of friends, observations elsewhere, trouble/poverty* and *no way out' without being creative* (Critchley *et al.*, 1999). The identity of farmer innovators as those who have developed or are testing new ways of land husbandry that combine production with conservation is well established. Such innovations may be simple cultural measures or sophisticated structural designs combined with integrated production systems. The innovation may be an on-going experiment, or already proven and effective. It may even have become established as a local tradition. As such, farmers' innovation is 'A practice started and later improved by a farmer on her/his own initiative (problem oriented), without any external influence at all' (Critchley *et al.*, 1999a). Indian farmers have also been continuously improving available technologies for more efficient and cost effective farming, which resulted in numerous innovations over the generations and helped in improving farming practices ensuring better livelihood options. Promotions of farmers' innovations are also in priority agenda of the government now a day in the form of various programmes.

Farmer-led innovations are evolving under specific agro-climatic and socio-economic conditions and such innovations should be widely adopted and sustained. The farmers identified a number of new/indigenous traditional crops and developed varieties with enhanced productivity and better quality through selection. Farmers also developed low cost processing technologies for value addition, increased shelf-life and better marketability for various farm products. In addition, a number of farm implements and tools were designed and manufactured by the farmers to increase operational efficiency and productivity. Women farmers have contributed in diverse germplasm conservation, postharvest management and value addition which helped in enhancing the farm income. Most of the farming practices traditionally adopted by the farmers are those which were evolved after long experiences of the farmers and communities under specific agro-climatic and socio-economic conditions.

Farmer -led innovations, over generations, have neither been duly recognized nor documented. Also the Intellectual Property Rights (IPR) on the innovations made by farmers has often been ignored. Value of traditional knowledge and its documentation has often remained unnoticed by scientists. The innovations led by farmers have neither been institutionalized for their horizontal and vertical expansion nor properly recognized. As a result, many technologies developed by innovative farmers have not reached to other farmers. Although, the initiative in the form of protection of propriety rights of the farmer-led innovations by government and non-government bodies have been taken in recent past but at a limited scale. Farm Innovators could have effectively become consultants and entrepreneurs leading to off- farm income generation options but the lack of proper sustenance support, institutional requirements for up scaling and out scaling and related constraints hindered farmer to farmer extension, institutionalization of such innovations and their blending with the modern scientific knowledge for the benefit of farming community at large.

Connotation of Farmer Led Innovation

In July 1987, some 50 social and natural scientists of roughly equal numbers met at the *Institute of Development Studies (IDS) at the University of Sussex*, UK, for a workshop on 'Farmers and Agricultural Research: Complementary Methods', later more generally known as the 'Farmer First' workshop. That workshop brought together experiences from a diverse range of individuals and organisations and marked a key moment in the development of approaches to farmer participation in agricultural research and extension. The importance of traditional knowledge for the protection of biodiversity and the achievement of sustainable development started slowly being recognized internationally (Gadgil *et al.*, 1993). The motivating factor for developing innovation has always been curiosity followed by increased production in addition to the reduced cost of production (Bayer, 2013). The sustainability of livelihoods becomes a function of how men and women use asset portfolios on both a short- and long-term basis (Krantz, 2001). SRI was a 'farmer-first' innovation from the outset, truly farmer-centered in the experimentation and evaluation undertaken by Fr. Henri de Laulanié (1993, 2003) and it has been farmer-participatory in its further development. This reporting on farmer innovation reflects that fact that SRI is still a work in progress. Chapters of this story have been written in over two dozen countries already, through the efforts of NGO workers, researchers, teachers, administrators, and other motivated individuals (Uphoff, 2009). Tripps (2006) study of low external input technology (LEIT) showed that there are relatively few examples where LEIT projects have led to a significant amount of independent experimentation, while farmers' experimental capacities got strengthened. Similarly, Sambodo (2007) mentioned that, farmers' decision rules can be distinguished according to their perceptions and attitudes, to their belief that they have the power to deal with problems and opportunities, as well as to the extent of socio-cultural influences. The innovators in general share certain distinctive characteristics like opportunism, tend to be *curious, proud and willing to take risks,* and they pick up ideas from here and there (CDCS, 1997), *they respond to recognition* (Gupta, 1998); they have latent skills and enthusiasm (Segeross, 1996), and triggered to innovate by various factors including *problem solving and accidental or even playful discoveries* (Roling, 1996) whereas the farm innovators *depend on the land; pick up piecemeal advice and blend with their own experience; focus on intensification and integration of resources; typically concerned with resource management; demonstrate pride in their own achievements, driven by a financial motive as well as a general concern with production, stimuli to innovation are travel outside the area, and information from various sources* (Critchley and Mosenene,1995).

The Continuum of Farmers' Innovation

The significance of farmers' innovation however ranges from being useful only to the individual farmer, to a wider range of farmers. Innovation involves the interaction of individuals and organizations possessing different types of knowledge within a particular social, political, policy, economic, and institutional context

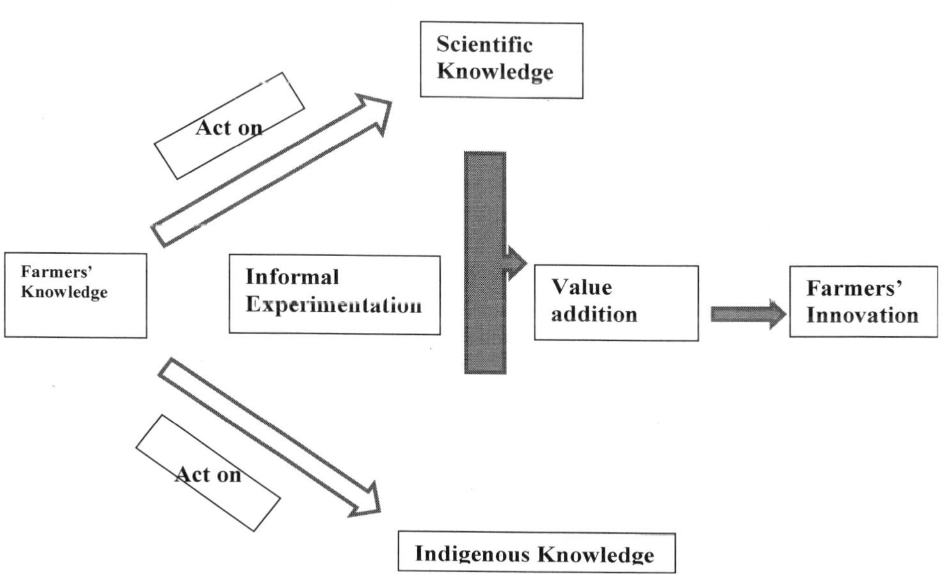

(World Bank 2006). Important dimension of the concept of "farmer innovation" is that it embraces not only technological innovation, but also new ways of managing livelihood in general (networking, communication, institution building, information management, marketing, planning, accessing resources, *etc.*). As a conceptual understanding platform, Prolinnova-Ethiopia (2004) adopted the following schematic presentation to explain the conceptual framework of farmer innovation:

Local innovations have been extended in different scales, however, 70 percent of local innovations were extended beyond the village and 30 per cent were confined only in village. The effectiveness of local innovation was reflected through its area of coverage. Farming community has developed a number of valuable innovations, and has proved potential through the application of these innovations in fellow farmers' fields also (ICAR, 2010). As far as the impact of farmer led innovation is concerned increased production has been the major outcome of most of the farmers' innovations followed by increased satisfaction and knowledge (Tambo, 2014). Another important area of impact of farmer led innovations may be farmers' capacity to continue the process of innovation to address other challenges through strengthening individual capacities, such as confidence, knowledge and to handle experimentation and innovation (Wettasinha *et al.*, 2014). Social networking of farm innovators has proved to be potential to construct knowledge. On the other hand the farm innovators require certain distinguishing capacities like foreseeing institutional requirements and linkages, comparative financial impact and success analysis ability in addition to analyse projected demand and required changes in socio cultural and infrastructural domain. FLIs having additional advantage over conventional innovations to tackle second generations' problems require different set of capacities on the part of farm innovators to scale their innovations in addition to be innovative, learning institutes for which are yet to be come into existence. (Nain *et al.*, 2018). Farmer-Led Agricultural Innovation for Resilience (FLAIR) has achieved tremendous success in introducing SRI to a large number of small-scale farmers. In Cambodia and Vietnam, SRI adopters experienced significant positive impacts on food security and income with a coarse calculation giving a 13-fold return on investment (Pommier, 2014). Innovative ways to increase production, improve organisation, or reduce dependence on external inputs, farmer innovations were found having significant potential to improve the quality of life for farming families in Malawi and reduce their impact on the environment. The farmer led innovations are driven by a range of interlinked factors: economic factors (the inability to afford external inputs or grow enough food to be food secure), environmental factors (the need to adapt to climate fluctuations or restore infertile soils which cannot be rested due to small landholdings), social factors(migration, HIV/AIDS, and scarce labour availability), cultural factors (need to use certain plants for ritual and other purposes), and political factors (availability of subsidized fertilizers and seeds as a form of political patronage by a neopatrimonial state (FAO, 2012)).

The Major Initiatives

The technological and institutional innovations are not two different departments of the same system, but it is often very common to see instances where technological innovations causing institutional changes or institutional innovation stimulating technological innovations. At international level, to develop mechanisms for local innovations to find their way into the formal research and development system, the Participatory Adaptation and Diffusion of Technologies for Rice-Based systems project initiated several activities to encourage their national partners to document, validate and disseminate local knowledge and innovations. This IFAD-funded project is coordinated by the African Rice Center (WARDA). The first phase of the project was implemented from 2000 to 2003 in Ghana, Guinea, The Gambia and cote d' Ivore. The Farm level Applied Research Methods for Eastern and South Africa(FARMESA) is a regional collaborative institute operating in five countries including Kenya, Tanzania, Uganda, Zambia and Zimbabwe with associate countries including Botswana, Malawi, Mozambique and South Africa (FAO. 2008). African Highlands Initiative (AHI) in traduced numerous technologies to improve and enhance land productivity in a sustainable way within the intensive land-use systems of the highlands in eastern and central Africa while maintaining the quality of the natural resource base during 1998-99, where farmers modified some elements of the technologies in different ways and sometimes opposing some of researchers methodologies in their fields according to their experiences with locally available alternatives resulting to a formal survey to trace such innovations and search out the motivating factors for innovation and its effects on adoption (Lyamchai *et al.*, 2005). The common activities for PROLINNOVA (Promoting Local Innovation) global programme are identification and documentation of local innovation, capacity building of different stakeholders, validation, promotion and up scaling of innovation which supports in livelihood of resource poor and low resource farmers. PROLLINNOVA Cambodia started up action research on so-called Local Innovation Support Funds (LISFs) in 2007 where farmers are given flexibility and independence in doing their own research relevant to local problems and conditions and facilitated the sharing of such innovations and experiences through a workshop. Promotion of farmer innovation and experimentation in Ethiopia (PROFIEET) Later named as PROLINNOVA-Ethiopia worked on the process and products of local innovation-based partnership between farmers, the formally trained researchers/ experts, the private sector, policy people, extension workers and other factors". The global community of practices included 16 countries mainly from Africa, Latin, America and Asia (PRROLINOVA-Ethiopia, 2006). Promoting Farmer Innovation (PFI) was the key activity of a project in East Africa upto 2001. The methodology and early results of PFI sought to use local farmer innovation in the field of land husbandry as a stimulus to more appropriate research and extension systems in semiarid and marginal areas which led to documentation of selected innovations as one of the products. International Assessment of Agricultural Knowledge, Science and Technology for Development (IAASTD) in northern Malawi stressed and worked for convergence between informal innovation and formal research and development systems (Olivia, 2012).

In Indian context, NIF (National Innovation foundation) took up the mission of making India innovative (documenting, adding value, protecting the intellectual property rights of the contemporary unaided technological innovators, as well as of outstanding traditional knowledge holders) and a creative society since 2000 with the active support of Department of Science and Technology, Government of India.

In this, it supplements the work done by the Honey Bee network over the years. The NIF seeks to create an innovation-driven society by giving awards to outstanding grassroots innovators, helping transform those innovations that have economic potential into products that can be commercially produced (either by the innovator himself or through licensing of the innovation to another commercial enterprise), and linking grassroots innovators to the formal science and technology system to get inputs to improve upon innovations wherever necessary and create a new model of innovation-driven entrepreneurship. To support individual innovators, the Technopreneur Promotion Programme (TePP) has been used to provide capital to several of the individual innovators. To support the commercialization activities of the NIF, the government created a Micro Venture and Innovation Fund of about $ 1million (to be operated by the NIF with the involvement of the Small Industries Development

Bank of India) in 2002 and these funds are being used to scale-up innovations and facilitate the creation of viable enterprises. The Society for Research and Initiatives for Sustainable Technologies and Institutions (SRISTI), a global NGO based in India was set up in 1993 to provide support to the honey bee network and to enhance global networking and do research on problems related to diffusion and protection of grassroots innovations. Grass Roots Innovation Augmentation Network (GIAN), Gujarat set up in 1997 is involved in documentation and validation of farmer innovations. It provides small amounts of funding for prototype development, links innovators to science and technology institutions, and identifies commercial enterprises interested in licensing product technologies from the grassroots innovators. ICAR organised a National farm innovators meet was organised at JSS Krishi Vigyan Kendra, Suttur, Mysore District, Karnataka and published Farm Innovators 2010. Intellectual Property and Technology Management cell (IP and TM) of ICAR oversees all matters related to intellectual properties and technology transfer/commercialization of new ideas. It is maintaining data base of successful innovative farmers for better dissemination. ICARs Zonal Technology Management and Business Planning and Development units (ZTM and BPD) main objective is to protect Intellectual Property Rights, showcasing, transferring and commercializing the ICAR institutes innovations. They also act as Agri Business incubator to Incubate new startup businesses. Almost all the State Agricultural Universities, state departments, KVKs, ATMAs are documenting the farmer led innovations at district level and recognizing them through kisan melas, exhibitions, seminars, conferences *etc.* (ICAR report 2015). Technology Information, Forecasting and Assessment Council (TIFAC) set up in 1988 under the Department of Science and Technology to look ahead in technologies, assess the technology trajectories, and support technology innovation by network actions in selected technology areas of national importance is providing the technical and financial support in the form of filing patents, extending post patent support for technology refinement and marketing and to upscale through various programmes. Protection of Plant Varieties and Farmers Rights Authority (PPV and FRA) set up in 2005, is involved in documentation, indexing and cataloguing of farmers' varieties. National Bank for Agriculture and Rural Development (NABARD)set up a separate fund titled "Farm Innovation and Promotion Fund" (FIPF) to encourage specifically the innovations in the farm sector. The Fund has been created in NABARD with an initial corpus of 5 crores and it was operational with effect from 1 April 2005. It was initiated mainly to provide support on the analogy of venture capital for innovative ideas – technological and managerial (like supply chain management) in farm sector for further development (NABARD report 2010).National Research Development Corporation (NRDC) was established in 1953 is recognizing a large repository of wide range of technologies spread over almost all areas of industries, *viz.* Agriculture and Agro-processing, Chemicals including Pesticides, Drugs and Pharmaceuticals, Bio Technology, Metallurgy, Electronics *etc.* It has licensed the indigenous technology to more than 4800 entrepreneurs and helped to establish a large number of small and medium scale industries. (NRDC annual report 2015)

Learning Exchange through Social Networking

The growing importance of knowledge has led to the concept of knowledge management which is the process of capturing, developing, sharing, and effectively using knowledge. Extension manages knowledge in an agricultural innovation system to support the progress of farmers. In general, the uses of information technology communications tools support knowledge sharing (Eid and Nuhu, 2011). The SECI (Socialization, Externalization, Combination, Internalization) model is a knowledge management model that explains how the different forms of knowledge are transferred or combined in an organisation. Social network or negotiation communication model considers information overload or fatigue which implies different approach for different issues and facilitative knowledge exchange (interactive partnership to share knowledge and experiences for decision making). Low-cost information and communication technology (ICT) tools possess the ability to deliver timely, relevant, and actionable information to farmers at lower costs than traditional extension services (Aker,2011, Cole and Fernando, 2012, World Bank, 2016). Social media has proved to be an important platform for pluralistic extension, bringing together all the actors in Agriculture Information System (AIS) and making them shareholders in development (Naruka *et al*,2017). Study on social interactions between online communities in online learning through mobile device, has found social presence to be a principal factor influencing motivations to engage in social interactions for constructing and sharing knowledge (Cheung *et al.*, 2008). Comparison of the use of blogs and Facebook for supporting knowledge management activities of creation, sharing, and application found that both tools generally support knowledge management but Facebook has more capabilities and potential than blogs in support of knowledge sharing (Cheung *et al.*, 2013).

Farmer led innovations (FLIs) having advantage to tackle second generations' problems require different set of capacities on the part of farm innovators to scale their innovations in addition to be innovative (Nain *et al.*, 2018). It can empower individual farmer and rural communities, strengthens link between farmers, extension worker and researcher in such a way that farmer experimentation directs the research agenda and the participatory technology development ensures sustainability of technology. Provision of comparative experiences through knowledge management systems, conflict management approaches, facilitation of multi stakeholder negotiations, building alliances with private sector, marketers and NGOs need to be stressed upon. Farmers need to initiate group action in production process, the mechanism for better remuneration need to be ensured for the extra efforts and the institutional arrangements for networking of stakeholders need to be devised to translate the challenges into opportunities.In order to enhance awareness of the innovative capacities of the farmers, to identify farmer-led innovations having potential to be adopted for larger impact and to share the experiences of farmers-led innovations in the field of agriculture and allied sector and to set the ground for networking of farm innovators, research institute and agricultural marketing agencies for dissemination of farmers' innovations as well as institutional innovations among larger population

Entrepreneurship Development through Farmer Led Innovation

There is proven nexus of entrepreneurship and innovation for sustainable development and need of the day is to encourage 'entrepreneurial agriculture for human development and maximum farm profits'. Innovativeness has been found to be critical in entrepreneurial behavior. Farm Innovators could effectively become consultants and entrepreneurs leading to off- farm income generation options after getting training and support in certain distinguishing capacities like foreseeing institutional requirements and linkages, comparative financial impact and success analysis ability in addition to analyze projected demand and required changes in socio cultural and infrastructural domain (Nain *et al.*, 2018). On other hand according to social network theory, entrepreneurs' social ties influence their recognition of entrepreneurial opportunities and entrepreneurial pursuits (Hills *et al.*, 1997). The development of a rural entrepreneurial support system necessitates creating a supportive environment, or social networking, to flourish in an entrepreneurial climate through building partnerships (Dabson *et al.*, 2003). Developing partnerships includes the coordinated efforts of central government, local governments, municipalities, academies and non-governmental organizations to help spur the entrepreneurial activity of that region (Kulawczuk, 1998). Partnership with institutions, academies and various organizations encourage rural community development and strengthen institutional support structures and well-built relations between the government and the private sector in new enterprise development in rural provinces. The strength of infrastructure development plays a crucial role in rural entrepreneurship development (FAO documents, 1997). Infrastructure development is highly correlated with the level of entrepreneurial activity across different countries (Zacharakis *et al.*, 1999). Since basic infrastructure development and availability of financing (Kulawczuk, 1998) are necessary for any entrepreneurial venture, it is assumed that a country's rate of the development of the national framework conditions may be a crucial link between a variety of other social, intellectual and environmental dimensions and rural opportunity recognition in a country. There is no proper appreciation of farmers as actors in the innovation system, little information provided about different sources of knowledge involved, or the flow of knowledge and little attention to long-term impacts on livelihoods (Brigidletty *et al.*, 2012). Institutionalization of any farmer led innovation being a complex process requires capacity strengthening and change in individuals as well as change in organizations. Fuentes *et al.* (2013) suggested that private players should assist in the commercialization of farmer-led innovations. Farmers should play a key role in planning the process of scaling out in their area to develop ownership and commitment to improving livelihoods. Supporting organizations need to facilitate the scaling out process beyond short term research or development projects (Miller and Connell, 2009).

IARI Experiences

In order to develop the model for making farming a business venture and replicating farmers' innovations, an action research study has been conducted and the farmer led innovations identified, documented for their scalability and the action interventions were initiated in three NCR Delhi villages namely *Fatehpur Biloch (Faridabad), Manjhawali (Faridabad) and Swamika (Palwal)* were selected purposively for action interventions being predominantly engaged in agriculture and having scope of agripreneurship development due to their proximity to National capital of Delhi. At first stage 135 farmers and farm women (45 from each village) were identified to analyse the perceived determinants for maximizing farm income and capacity building needs in agripreneurship development. On the basis of need analysis action interventions were identified and at this stage 110 farmers and farm women were involved on the basis of their interests and motivations. Pre training and post training data on entrepreneurial competencies as suggested by McClelland, 1969 was collected from 110 farmers and farm women trained on various aspects of entrepreneurship development. In order to understand the backward and forward linkages, 30 farmers each (a total of 120 farmers) cultivating tomato, cauliflower, tuberose and gladiolus were interviewed and the perceived linkages were mapped on three point continuum from poor linkage, fair and good linkages. The price spread (the difference between the price received by the growers and the price paid by the consumers)for four farm products namely tuberose, gladiolus, cauliflower and tomato was calculated with standard procedures and the estimate of producer's share in consumer rupee was performed. To test the scalability of the farmer led innovations, a test was standardized consisting of seven broad parameters namely; credibility, complexity, testability, observability of results, relevancy, relative advantage over existing practices and sustainable source of funding with suitable modifications in scaling up toolkit of Cooley and Ved (2012). The data for analysis of scalability of the innovation were collected from 60farmers (20 from each village) from project locations. Simple statistical tools averages, percentage, mean score, weighted mean score were employed to accomplish the different objectives of the study.

On the basis of analysis of the successful cases and the action interventions undertaken a framework for agri-entrepreneurship and farmers' innovation dynamics has been conceptualized. The agri-entreprise development for maximizing farm profitability was found to be interplay of entrepreneurial competencies, entrepreneurial climate, and farmers' innovations. It was found that the competencies like opportunity recognition, drive for excellence, quality concern,moderate risk taking behaviour, innovativeness and business orientation in presence of suitable climate like networking, infrastructure, government priority and financial backstopping lead to experimentation not only for technological innovation, but also new ways of managing livelihood in general (networking, communication, institution building, information management, marketing, planning, accessing resources, *etc.*). The innovations which were economically viable and found sustainable source of funding were able to translate into entrepreneurial ventures having higher income and profits. Social networking of farm innovators has proved to be potential to construct knowledge. On the other hand to maximize the income, the farms required certain distinguishing capacities like foreseeing institutional requirements and linkages, comparative financial impact and success analysis ability in addition to analyse projected demand and required changes in socio cultural and infrastructural domain. The results are in conformity with Singh *et al.*, 2014 and Singh *et al.*, 2016, whereas it was inferred that individual motivations and aspirations trigger entrepreneurship and the competencies along with best practices (innovations) and convergence and synergistic linkage play sequential role for enterprise success.

Summing Up

Diminishing returns from Agriculture have put the whole rural economy under severe stress which paved the way to accelerate the diversification of rural livelihoods in order to have adequate viable and sustainable means and opportunities. Institutional mechanism and human mobilization for networking and resource optimization, collectivization, technologies and methodologies of secondary agriculture are the keys for maximizing farm income. Human resources base in rural ecosystem in general is lacking in social processes of group and enterprise management skills along with marketing and communication skills. The capacity building interventions not only have the potential for changing entrepreneurial competencies but broadening the horizon of the participants to launch their own income generating activities. The backward and forward linkages in the form of advisory services, input supply, marketing of the produce, financial backstopping and the support and convergence of various stakeholders like banks, NGOs, research institution, state line department may bring positive impact in the form of initiation of income generating activities out of their own innovations. Screening for scalability of farmers' innovations and efforts for their institutionalization pre- requires creation of platform for exchange of information and experiences, developing and disseminating theme-based knowledge products and undertake analysis of partner institutions to assess their potential as participants and building capacity of partner institutions. The framework for agri-entreprise development for maximizing farm profitability may be conceived as the function of entrepreneurial competencies, entrepreneurial climate, and farmers' innovations. To encourage Entrepreneurial Agriculture for Human Development and maximum farm profits, farmers need to harness all their skills so that they will be able to withstand harsh conditions which are as a result of environmental changes or social compulsion driven. Agricultural productivity is believed to be enhanced by the incorporation of strategic entrepreneurship skills. Also, farmer led innovations generated for immediate problem solving or creative application have helped in optimizing farm profits and managing agricultural activities conveniently. Innovativeness has been found to be critical in entrepreneurial behaviour.

REFERENCES

Aker, J.C. 2011. Dial "A" for agriculture: a review of information and communication Amanuel Assefa.2005. Challenges and prospects of Farmer Innovation In Ethiopia Presented in the workshop organized by IFAD under the theme "what are the innovation challenges for rural development" on 15-17 Nov 2005, Italy, Rome

Bayer, S. 2013. What determines innovation capacity in farm households? Insights from rural Ghana. [*Retrieved from https://editorialexpress.com/cgibin/conference/download.cgi? db_name=CSAE2014 and paper_id=391*].

Bhattarai, S.M., Dangol, D.R., Srivastav S.B. and Shrestha P.K. 2015. Factors influencing local innovation in ecological agriculture in the central development region of Nepal. *Nepalese Journal of Agricultural Sciences*, 15: 69-83.

Brigidletty, Zanele Shezi and Maxwell Mudhara. 2012. Agricultural grassroots innovation in South Africa: Implications for indicator development. Ideas for New Research Projects on LICS in Africa. African Globelics Seminar, Tanzania.

Chamber, R., Pacey, A. and Thrupp, L.A. (eds). 1989. Farmer First: Farmer, Innovation and Agricultural Research, Intermediate Technology Publication, London.

Chang, R.C.H., Chu, S.K.W., Lee, C.W.Y., Chan, B.K.T. and Leung, C.K. 2013. Knowledge Management using Social Media: A Comparative Study between Blogs and Facebook. Proceedings of the *American Society for Information Science and Technology* ASIST. 50(1): 1-9.

Cheung, W.S., Hew, K.F. and Ng, S.L. 2008. Toward an understanding of why students contribute in asynchronous online discussions. *Journal of Educational Computing Research.* 38(1): 29–50.

Cole, S.A. and Fernando, A.N. 2012. The value of advice: Evidence from mobile phone based agricultural extension. Working Paper, Harvard Business School, 13-47.

Critchley Will, Cooke Roshan, Jallow Tijan, Lafleur Sophie, LamanMineke, Njoroge Janet, Nyagah Verity and Obas Emmanuelle. 1999. Promoting Farmer Innovation harnessing local environmental knowledge in East Africa. Workshop Report no 2, UNDP – Office to Combat Desertification and Drought (UNSO/SEED/BDP), and Sida's Regional Land Management Unit (RELMA).

Critchley, W.R.S., Miiro, H.D. and Ellis-Jones, J., Briggs, S. and Tumuhairwe, J., 1999. Traditions and Innovation in Land Husbandry. Building on Local Knowledge in Kabale, Uganda. RELMA Technical Report no 20, Nairobi, Kenya.

Critchley, W.R.S., Mosenene L. 1996. Individuals with initiative: network farmers in Lesotho. In Successful Natural Resource Management in Southern Africa, Centre for Development Cooperation Services, Vrije Universiteit Amsterdam (ed.). Gamsberg Macmillan: Windhoek; 71–81.

Dabson, B., Malkin, J., Matthews, A., Pate, K. and Stickle, S. 2003. Mapping Rural Entrepreneurship. Battle Creek, MI: W. K. Kellogg Foundation; and Washington, DC: CFED.

Eid, M. and Nuhu, N.A. 2011. Impact of learning culture and information technology use on knowledge sharing of Saudi students. *Knowledge Management Research and Practice.* 9(1): 48-57.

FAO. 1997. Higher agricultural education and opportunities in rural development for women – An overview and summary of five case studies. M. Karl. Rome.

FAO. 2012. Recognising the Unrecognised: Farmer Innovation In Northern Malawi A Report by Find Your Feet. [*Retrieved from http://www.fao.org/fileadmin/templates/agphome/Family_farming/Farmer_Innovation_Study_Final.pdf*]

Fuentes and Ernst Berg. 2013. Impact Assessment of Agricultural Innovations: A Review. *Agronomia Colombiana,* 31(1): 120–30.

Gadgil, M., Berkes, F., Folke, C. 1993. Indigenous knowledge for biodiversity conservation. Ambio, 22: 151–156.

GIAN - www.gian.org.

Gubbels, P. 1997. Strengthening community capacity for sustainable agriculture. *In* van Veldhuizen L, Waters-Bayer A, Ramirez R, Johnson DA and Thompson J, eds. *Farmers' Research in Practice: Lessons from the Field.* London: Intermediate Technology Publications. 217–44.

Gupta, A. 1989. Scientists' view of farmers' practices in India: Barriers to effective interaction. In R. Chambers, A. Pacey, L. A. Thrupp (eds.) Farmer First. Farmer Innovation and Agricultural Research (pp. 24–31). London, UK: Intermediate Technology Publications

Gupta, A. 1998. Respecting recognizing and rewarding local creativity: Knowledge networks for biodiversity conservation and natural resource management. Mimeo. Ahmedabad, India: Indian Institute of Management.

Gupta, A. 2000. Grassroots Innovations for Survival. ILEIA Newsletter for Low External Input and Sustainable Agriculture (LEISA), 16(2): 5–6.

Hall, A. and Dijkman, J., 2006. Capacity development for agricultural biotechnology in developing countries: concepts, contexts, case studies and operational challenges of a systems perspective. UNU-MERIT Working paper series No. 003, United Nations University, Maastricht Economic and social Research and training centre on Innovation and Technology. [*Retrieved from http://www.merit.unu.edu/publications/wppdf/2006/wp2006-003.pdf*]

Hills, G.E., Lumpkin, G.T., and Singh, R.P. 1997. Opportunity recognition: Perceptions and behaviors of entrepreneurs. *Frontiers Of Entrepreneurship Research.* 17(4): 168- 182.

ICAR report 2015. https://icar.org.in/content/annual-report-2014-2015

ICAR. 2010. Farm Innovators. New Delhi: Division of Agricultural Extension, Indian Council of Agricultural Research, New Delhi.

IDS Workshop 1989. Farmers' knowledge, innovations, and relation to science. In Chambers R., Pacey A., Thrupp L. A. (eds) Farmer First. Farmer Innovation and Agricultural Research (pp. 31–38). London, UK: Intermediate Technology Publications

ILEIA 2000. "Grassroots innovation. Unleashing the creativity of farmers" *Newsletter for Low External Input and Sustainable Agriculture* (*LEISA*).16(2): 4–5.

Joanne, M. and Connell John. 2010. Strategies for scaling out impacts from agricultural systems change: the case of forages and livestock production in Laos. *Agriculture and Human Values*, 27: 213–25.

Krantz, L. 2001. The Sustainable Livelihood Approach to Poverty Reduction: An Introduction. Swedish International Development Cooperation Agency. Division for Policy and Socio-Economic Analysis.

Kulawczuk, P. 1998. The development of entrepreneurship in rural areas. *In* J. D. Kimball (Ed.), The transfer of power: Decentralization in central and eastern Europe (pp. 97–109). Budapest: The Local Government and Service Form Initiative.

Laulanié, H. (2003). Le Riz à Madagascar: Undèveloppementen dialogue avec les paysans, Editions Karthala, Paris.

Laulanié, H. 1993. Le système de riziculture intensive malgache. *Tropicultura.* 11:110-114.

Laulanie, H. 2003. Le Riz a Madagascar: Undevelop pementen dialogue avec les paysans, Editions Karthala, Paris.

Laulanié, Henri de. 1993. Le système deriziculture intensive malgache. Tropicultura (Brussels), 11(3): 110-114.

Lyamchai, C.J., Kingamko, M.N., Mowo, J.G and German, L. 2005. Farmers' innovations in natural resource management: lessons and challenges from lushoto, *International Center for Research in Agroforestry.* Tanzania.

NABARD report 2010. https://www.nabard.org/auth/writereaddata/tender/3107173739SMFI per cent 202009-10 per cent 20Eng.pdf

Naidu, Y.D. and van Walsum, E. 2003. PTD for sustainable dryland agriculture in South India: Balancing our way to scale. *In* Wettasinha C, van Veldhuizen L and Waters-Bayer A, eds. *Advancing Participatory Technology Development.* Silang, Philippines: IIRR. 215–36.

Naidu, Y.D., Lanting, H. and Kolli, R.D. 1998. Leaf wetness counter: A case study of institutional partnership towards sustainable groundnut production in South India. Paper presented at International Workshop on NGO–Research Partnerships, 4–10 October, Silang, IIRR.

Nain, M.S., Singh, Rashmi and Mishra, J.R. 2019. Social Networking of Innovative Farmers through WhatsApp messenger for Learning Exchange: A study of content sharing. *Indian Journal of Agricultural Sciences.* 89(3): 556-558.

Nain, M.S., Singh, Rashmi, Mishra, J.R. and Sharma, J.P. 2015. Utilization and Linkage with Agricultural Information Sources: A Study of Palwal District of Haryana State. *Journal of Community Mobilization and Sustainable Development.* 10(2):152-156.

Nain, M.S., Singh, Rashmi, Mishra, J.R. and Sharma, J.P. 2018. Scalability of farmer led innovations (FLIs): A study of perceived determinants and required capacities. *Indian Journal of Agricultural Sciences.* 88(8): 1312-1315.

Nain, M.S., Singh, Rashmi, Sharma, J.P., Burman, R.R. and Chahal, V.P. 2015. Participatory identification and prioritization of agri enterprises in national capital region of India. *Indian Journal of Agricultural Science.* 85(6): 787-791.

Naruka, P.S., Verma, Shilpi, Sarangdevot, S.S., Pachauri, C.P., Kerketta, Shilpi and Singh, J.P.2017. A Study on Role of WhatsApp in Agriculture Value Chains. *Asian Journal of Agricultural Extension, Economics and Sociolog.* 20(1): 1-11.

NIF. 2004. National Innovation Foundation. *NIF Update.* September. NIF- www.nifindia.org.

Norman, U. 2009. Farmer Innovations Improving the System of Rice Intensification (SRI) In Scoones, I and Thompson, J (Eds) *Farmer First Revisited: Innovation for Agricultural Research and Development,* Oxford: ITDG Publishing.

NRDC Annual Report. 2015. https://www.nrdc.org/sites/default/files/nrdc-annual-report-2015.pdf

Olivia, Wills. 2012. Recognising the unrecognised: Farmer innovation in northern Malawi. A report by Find Your Feet.

Pommier Ludovic. 2014. Evaluation of farmer led agricultural innovation for resilience 2010-2012. OXFAM AMERICA Evaluation Report. [*Retrieved from https://www.oxfamamerica.org/static/media/files/Evaluation_of_the_Farmer-Led_Agricultural_Innovation_for_Resilience_2010-2012_.pdf*]

Profieet/Prolinnova. 2004. Proceeding of the First International PROLINNOVA Workshop, 8–12 March 2004, Yirgalem, Ethiopia. Addis Ababa: ASE/Silang, Philippines: International Institute of Rural Reconstruction.

Prolinnova- Ethiopia Progress Report. 2006. Unpublished report

Reij, C. and Waters-Bayer, A. (eds) 2002. Farmer Innovation in Africa. A source of inspiration for Agricultural Development. Earthscan, London.

Röling, N.G. 1996. Towards and Interactive Agricultural Science. *European Journal of Agricultural Education and Extension,* 2: 35– 48.

Sambodo, L. 2007. The decision making processes of semi-commercial farmers: A case study of technology adoption in Indonesia. Ph.D. Thesis, Lincoln University, New Zealand.

Segeross, M., Prasad, G. and Marake, M. 1996. Let the farmer speak: Innovative rural action learning areas. In: CDCS (ed.), Successful Natural Resource Management in Southern Africa. Windhoek, Namibia: Gamsberg Macmillan Publishers Ltd. 91-116.

Singh, R., Nain, M.S., Sharma, J.P. and Mishra, J.R. 2016. Developing Agripreneurship for Sustainable Farm Income: Action Research Study on Women Farmers of Hapur District, Uttar Pradesh. *Journal of Community Mobilization and Sustainable Development.* 11(1):127-135.

Singh, Rashmi, Nain, M.S., Sharma, J.P., Mishra, J.R. and Burman, R.R. 2014. Institutional Convergence of Synergistic Strengths for Developing Women Agripreneurs. *Indian Journal of Extension Education.* 50(3 and 4): 1-7.

Tambo, A. J. 2014. More than adopters: the welfare impacts of farmer innovation in rural Ghana. Selected Paper prepared for presentation at the Agricultural and Applied Economics Association's, AAEA Annual Meeting, Minneapolis.

Tripp, R. 2006. Self-Sufficient Agriculture. Labour and Knowledge in Small-Scale Farming. London: Earthscan.

Wettasinha, C., Waters-Bayer, A., van Veldhuizen, L., Quiroga, G. and Swaans, K. 2014. Study on impacts of farmer-led research supported by civil society organizations. Penang, Malaysia: CGIAR Research Program on Aquatic Agricultural Systems. Working Paper. AAS-2014-40.

World Bank. 2006. *Enhancing Agricultural Innovation: How to Go Beyond the Strengthening of Research Systems.* Washington, DC: World Bank.

Zacharakis, A.L., Reynolds, P.D., and Bygrave, W.D. 1999.Global Entrepreneurship Monitor: National Entrepreneurship Assessment for the United States of America. Kansas City, KS: Kauffman Center for Entrepreneurial Leadership.

www.africarice.org/warda/aboutus.asp

www.prolinnova.net/Nepal

www.sristi.org/hbnew

Chapter 6

Information and Communication Technology for Effective Extension Service Delivery and Sustainable Development

R.N. Padaria, Shantanu Rakshit, S. Aiswarya and Priti Priyadarshni

Division of Agricultural Extension, ICAR-Indian Agricultural Research Institute, New Delhi – 110 012

Introduction

Agricultural advisory systems have changed over the past decades, especially with regard to public extension, as farmers receive information from a wide range of ICT sources. These changes in ICT during the past two decades have been rapid and far-reaching, from the costly, colossal, energy-consuming equipment once available to the very few towards the booming and continuously changing mobile, wireless, and internet industries for the many. In the last two decades, public and private actors operating in rural contexts in developing countries have widened their array of methods for disseminating knowledge through radio programmes, videos and other ICT-based methods.

ICT has the potential to respond to a number of challenges that confront public extension systems. The public extension system faces difficulty in reaching all farmers due to the lack of financial capacity and staff. This is exacerbated by the fact that farmers are sparsely populated across large areas and often isolated. Strong public extension services only directly reach about 10 per cent of the farmer population, and this is even less if operating funds are limited (Bell, 2015). Another key challenge is that farmers increasingly request specified and varied information, as farming is becoming more and more market-orientated. Even farming systems in relatively homogenous agricultural areas differ in type of crops cultivated, inputs, labour, machinery and quality standards used. In order to be truly effective, information and knowledge have to be tailored to meet the needs of the individual farmer.

Another limitation of public extension services is the non-recurring character of information and knowledge provision to farmers. In the most optimal case, the information delivered is updated and renewed by means of regular exchanges between extension officers and farmers. However, the disadvantage of this is that it is costly and time consuming. The absence of a local facilitator or expert means that these efforts do not always result in adoption by the farmer (FAO, 2015).

Finally, public extension systems in developing countries are known for being bureaucratic and lacking transparency and accountability. Underpaid extension officers might use their working time other than visiting isolated farmers. Support effectiveness of extension is often low, as farmers are often resource-poor and illiterate, the number of extension staff is limited, extension workers may lack relevant knowledge and skills, and also the institutional motivation and resources to reach farmers are low (Bell, 2015).

Status and Evolution of Information and Communication Technology in India

A report by Telecom Regulatory Authority of India (TRAI) revealed that the total number of telephone subscribers in India reached 1191.81 million as on 31 August 2019. The number of wireless subscribers are 1171.00 million and the number of wireline subscribers are 20.82 million. The numbers of internet users are 665.31 million, out of which 70.72 million are narrowband subscribers and 594.38 million are broadband subscribers. India has the world's second highest number of internet users after China, with 451 million users connected at end of financial year 2019. In India, the majority of internet users are found in urban areas, as there is relatively less access in the rural areas of the country. Like China, the primary link between the population and the internet is mobile smartphones. The report said that of the 451 million monthly active users, 385 million are over 12 years of age and 66 million are in the age bracket of 5 to 11 years, who access Internet on the devices of family members (TRAI, 2019). The report stated that in terms of absolute numbers, urban India with 192 million users had almost the same number of users as rural India. However, in terms of percentages or penetration, given the disparity of population distribution in urban and rural India, urban India had a considerably higher penetration level. In rural India, a sizeable portion does not have access to Internet, and provides a huge opportunity for growth, which will contribute to an increase in the overall Internet population over the next few years.

According to the report, nearly 2/3rd of the Internet population in India were daily users. About 9 out of 10 users' in urban accessed the Internet at least once a week with young users aged between 16-29 years being the most frequent users.

However, there is still a section of users who used Internet less than 'once a week'. About 1 in 5 rural Internet users belonged to this category. In India, nearly 1/3rd of users accessed Internet for 'more than one hour' in urban India, whereas in rural India, a similar proportion of users accessed Internet for 15-30 minutes. With better connectivity, quality of service and affordability of mobile Internet, there could be an increase in rural consumers spending more time on the internet in future, the report stated (TRAI,2019).

Major ICT related initiatives which were initiated during the latter part of the 20th century and accelerated during the initial part of the 21st century pertaining to agriculture in India are: a AQUA (Almost All Questions Answered), Digital Green– Participatory video for Agricultural Extension; e-Arik (e-agriculture); e-Sagu (e-cultivation), KISSAN (Karshaka Information Systems Services and Net Working); Life Lines India- Soochna Se Samadhan(Solutions Through Information);VASAT-Virtual Academy for the Semi-Arid Tropics, touch screen kiosks; MSSRF-Village Knowledge Centres; ISRO-Village Resource Centres; Community Information Centres (CICs); Common Service Centres (CSCs); Warna Wired Village Project; Various web based portals such as AGRISNET (Agriculture Resources Information System and networking); DACNET (Department of Agriculture and Cooperation and Network); INDG-India Development Gateway Portal; DEAL-Digital Ecosystem for Agriculture and Livelihood; i*KISAN* of Nagrajuna Group; e-*Krishi*; ASHA; IFFCO-Agri portal; Agri watch portal; i- Shakti; AGMARKNET; e-*KRISHI VIPANAN*; ITC-e-Choupal; EID Parry-India agri line; Kisan Call Centres (Farmer call centre); SMS Broad Cast Service by KVK; Uttam krishi.com and hello krishi by CHAMBAL Fertilisers; Airtel- IFFCO Kisan Sanchar Limited; BSNL-Mandi on mobile service; Nokia Life Tools; Fisher Friend Project; Help Line Service by the Chandra Shekar Azad University, Aswini Project; Community Information Centres; Digital mandi Project; e-krishiVigyan Project; KisanSoochana Kendra; Rural Knowledge Centre; SKS Micro finance; TANWA;MSSRF-Information Village Project; e-Krishi, *etc.* Apart from this, there are presently 185 Community Radio stations (*MIB*, 2019) which are in operation in various parts of the country to provide efficient services to the farmers and rural population.

Agricultural Extension and Advisory Services

Traditional extension services face several challenges in developing countries that limit their efficiency. Poor infrastructure makes it harder and more costly to visit remote areas. For this reason, often extension programmes provide only one-time information to farmers, lessening their long-term impact. In addition, traditional extension is plagued by principal-agent and institutional problems, including a lack of accountability. ICTs can increase smallholder's access to timely extension information while addressing many of these challenges by reducing the cost of extension visits, enabling more frequent two-way communication between farmers and agents, and improving agents' accountability. ICTs also enhance access to private information from social networks, thus facilitating learning from one's peers, which is crucial for technology adoption. By increasing communication linkages between farmers, extension agents, and research centers, ICTs can improve the flow of relevant information among all these agents.

Digital Green, a non-profit international development organization, uses an innovative digital platform for community engagement to improve the lives of rural communities across South Asia and Sub-Saharan Africa. Partnering with local public, private and civil society organizations the platform shares knowledge on improved agricultural practices, livelihoods, health, and nutrition, using locally produced videos and human mediated dissemination.

Analysis of the role of ICTs in agricultural extension highlights the heterogeneity of delivery systems, which include one-way versus two-way communication between farmers and agricultural specialists, SMS versus voice messaging, and oral description of problems versus pictures taken in the field. Although these applications increase farmers' awareness, this does not automatically translate into behavioral changes such as increased adoption of improved agronomic practices or modern inputs. Overall, there remains however a lack of evidence regarding which services work and which do not, as most agricultural extension through ICTs is fairly recent

Emerging Practices to Improve the Delivery of Information through ICT

Presently, a wide variety of approaches are being tried to use ICT to enhance extension services. Approaches differ in the format and how the information is transferred; for example, by text, voice or picture, and through queries or SMS messages. In some cases, there is direct communication between the sender and the farmer. In others, the messages are disseminated through a farmer extension worker or a local facilitator that has access to the Internet. Some approaches support two-way interaction enabling the farmer to respond to or request from the provider.

ICT-based extension advisory methods are relevant in areas such as pre-production, production, post-harvest and marketing, financial services, and gathering and distributing of data. Different tools are suitable for different applications (Saravanan *et al.*, 2015). In general TV, radio and video are used for awareness creation and transfer of technology. Mobile phones are mainly utilised for collecting and disseminating advisory and market information, such as prices and location. Web portals provide unique opportunities for information sharing and linking with other stakeholders and e-learning is specifically interesting for educational purpose. Social media integrates all functions; from providing advice and sharing knowledge to creating awareness, linking with other actors, and technology transfer. Community radio, telecentres, videos, virtual communities of practice and social media enable farmers and others to 'gain a voice'. Also, many extension interventions combine ICT channels such as mobile phone services with traditional communication channels, like radio (USAID, 2010).

Principles and Process to Strengthen the Delivery of Information through ICT

To successfully improve ICT-based extension systems, several conditions have to be met. First of all, the success of ICT is dependent on the knowledge of people on how to use devices and navigate the Internet. For example, hosting web portals and e-learning platforms requires advanced technical knowledge and computer skills. The same applies for app-development.

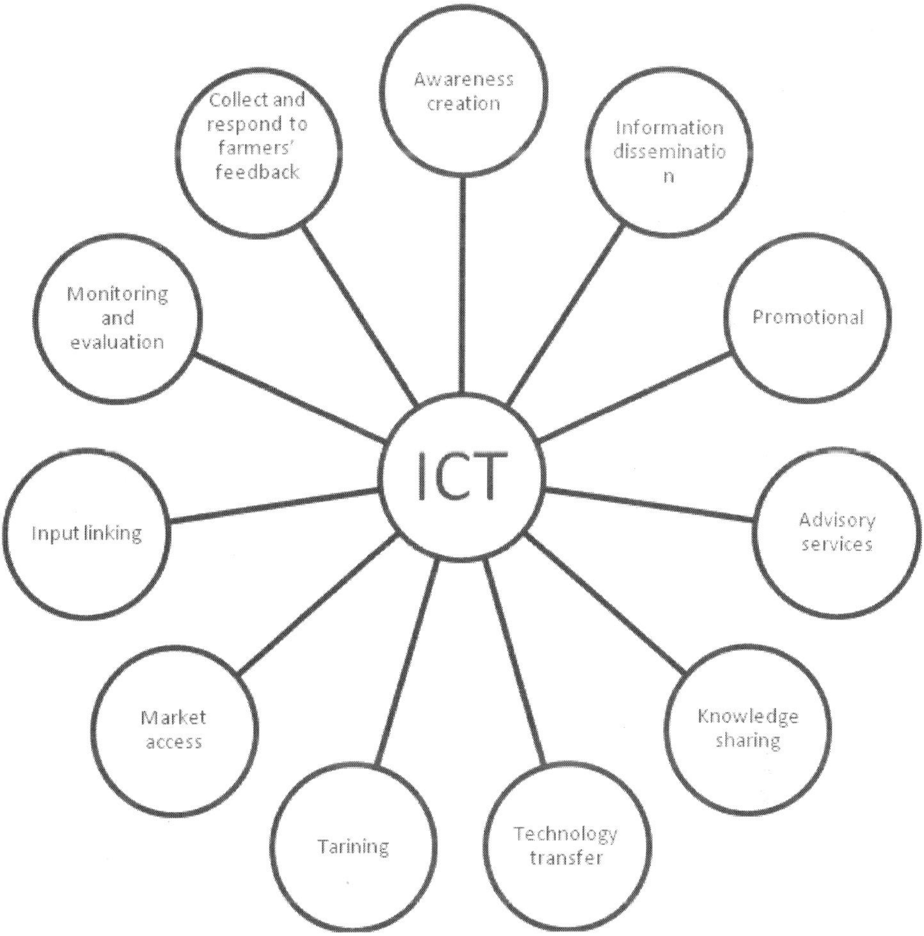

Functions of ICT Yools
(*Source*: Saravanan *et al.*, 2015).

Second, ICT only achieves impact when the mode used corresponds to the interest and capacity of the user group. Web portals, e-learning and text-based SMS messages are only useful for literate farmers, whereas video, voice based advisory services and community radio are more suitable for illiterate people. The tools selected have to match the purpose, content and clientele. Moreover, integrating traditional media and new ICTs can expand the reach of extension, but a high adoption rate requires farmers to be engaged in determining relevance and developing content and allowed to interact with information/service providers (Francis and Addom, 2014).

Third, it is important to realise that ICT does not generate content but acts as a vehicle to disseminate it. People make use of the services provided only when the content is of interest to them. This is more probable when services are timely, specific, contextualised and targeted. Moreover, ICT based services alone are not enough. They create greater synergy when combined with other extension methods like farmer field schools and demonstrations. To achieve widespread impact, institutionalising ICTs is therefore necessary (Christoplos, 2010). There has been an assumption that with the 'right' technological investment, extension agencies will achieve new objectives and become more sustainable. However, innovations within ICT must be adapted to prevalent constraints, such as institutional structures, human and financial capital constraints, in order to be scaled up and successfully implemented (Christoplos, 2010).

For ICTs to successfully enhance extension, Saravanan *et al.* (2015) suggested the following steps for implementation. Note that each step depends on the situation and judgement of the extension organisation:

1. The first action of the organisation/ministry/agency that is going to use ICT-based extension should be an assessment of the needs of the target community.
2. To adapt, monitor and evaluate ICT enabled services, it is important to conduct benchmark surveys before introduction. Benchmark surveys also help to get a good overview of the actual situation.
3. Based on the needs assessment and benchmark surveys, localised and customised content needs to be developed.
4. ICT tools need to be selected and developed in such a way that they correspond to the desires and needs of the target group.
5. The target audience should be sensitised on the presence of the services and how to access them.
6. The newly developed ICT-based services are introduced and used in extension.
7. To ensure the sustainability of the services, it is recommended to search for partnerships with stakeholders present in the target area or seek for integration of the services in the public agricultural extension system.
8. Monitoring and adaptation are important, especially at the beginning of the project. Modifications should be made when the project does not correspond to the needs of the audience.
9. Finally, an impact assessment should be realised to determine the degree of success of the project.

Use of ICT for Achieving Climate Change Resilience

The increasing use of information and communication technologies (ICTs) is creating new opportunities and challenges for developing countries that are at the forefront of climate change impacts. The ICTs, such as community radios, mobile phones, knowledge centres and interactive media, *etc.* help to reduce climate change vulnerability and risk while including the voices of those most affected group. ICTs support the process of climate change mitigation/adaptation through providing access to relevant information, raising awareness at the grassroots level, facilitate networking, and facilitating learning and practical knowledge sharing at the community level (Kalas and Finlay, 2009). The potential of information and communication technologies (ICTs) has not yet been integrated and utilised into a systematic understanding of adaptation and strengthen resilience. The ICTs can become strategic enablers of action to facilitate climate change awareness, mitigation, adaptation and monitoring (Ospina and Heeks, 2012). The use of geographic information systems (GIS) and remote sensing technologies can enable analysis of the spatial relationships between carbon emissions and local socioeconomic conditions, contributing to decision making processes and to the implementation of mitigation mechanisms. Applications such as technologies can help to the analysis of forest carbon stocks, as well as to the participatory management of forest resources within rural and marginalised settings (*e.g.* forestry information systems, geo-referenced databases of land tenure, and local awareness regarding reforestation/afforestation trends) (Ospina and Heeks, 2012). Climate monitoring can be strengthened through the use of ICTs for data capture, processing and dissemination by and among local stakeholders. ICTs are being increasingly used to map, record and analyse changes in natural resources such as water basins, animal and vegetal species and biodiversity, as well as changes in pollution levels and greenhouse gas emissions, fostering the engagement of community actors in participatory monitoring and information exchange. ICT-based tools such as internet portals and online databases are used to track the allocation and disbursement of climate change related funds, contributing to transparency and accountability.

ICTs can strengthen the preparedness for climate change related events through applications such as geographic information systems (GIS), and positioning and modelling applications. ICTs can strengthen social networks through enhanced communication within those networks. ICT can facilitate access to a broader set of physical, financial, human and social capital, fostering the adaptive capacity of livelihood systems. ICTs enabled end users to interact with experts and government officials even located in different locations (Nanda and Arunachalam, 2009). The *Gramin Krishi Mausam Seva* (GKMS) of India Meteorological Department (IMD) is successfully providing the crop specific advisories to the farmers at the district level through different print/visual/Radio/IT based media including Short Message Service (SMS) and Interactive Voice Response Service (IVRS) facilitating for appropriate field level actions. Weather forecast based agro-meteorological advisories are disseminated through Kisan portal launched by the Ministry of Agriculture and also under Public Private Partnership.

Challenges for Promoting ICT in Extension

Despite the promise and potential of ICT in extension services, there are numerous challenges facing ICT as an extension strategy, which can be seen in the list below:

1. One key challenge is the scaling up of ICT in extension services. Many ICT interventions fail to scale up and achieve widespread adoption due to market fragmentation and the lack of financially sustainable business models that will attract private sector investments in innovative solutions for small-scale agriculture (World Bank, 2016). Rather than assuming that an ICT approach will always be cost-effective and yield a better outcome, there is a need for a more nuanced understanding of the underlying institutional environment and constraints (World Bank, 2016).
2. Complex and dense information, such as nuanced information on agricultural practices and inputs, have to be converted into ICT-based messages (Aker, 2011). To transform these into SMS messages or videos requires the capacity to summarise complex information into concise key points, which can be easily understood and utilised.
3. Whereas the expansion of mobile phone access has been rapid and commercially self-sustaining, even among many of the poor, the same does not hold for the internet. In the long run, however, the internet can have an even greater impact on rural growth. Much depends on finding sustainable business models that encourage its spread in the poorest parts of the world (World Bank, 2016).
4. Even with the information provided from the ICT intervention, it is not guaranteed that the farmer will act upon this information because of the inaccessibility of alternative markets and the complex interlinked relationships between buyers and sellers in low-income developing economies (World Bank, 2016).
5. Communicating information in the national language might lead to misunderstanding or low adoption rates, as in many rural areas farmers only speak local languages. In regard to adoption rates, integrating traditional media and new ICTs can expand the reach of extension. However, to achieve high rates of adoption, farmers need to be engaged in determining the relevance and content that will be shared and distributed to other farmers (Francis and Addom, 2014). This also relates to the capacity of the farmer to search information. Even if the information is simple and the message is fast and timely, reception depends on farmers' understanding of the phone, computer or any other medium used. This holds also for the extension worker or call centre agent sending the information; in case farmers respond with questions he or she must be able to search the answers and respond in an understandable way (USAID, 2010).
6. Another challenge is the diversity of farmers. Having the potential to tailor and personalise messages, ICT initiatives can reach a diverse and large number of farmers. However, this requires good knowledge of the user group. To make sure that not only the entrepreneurial young farmers receive the information it is important to differentiate between the farmers in a community. For example, women and elderly often have lower literacy and schooling rates than young men. This requires the

development and use of adapted information and training materials. In this respect, audio-visual tools are more likely to offer opportunities to reach women farmers (Quisumbing and Pandolfelli, 2009).

7. Finally, like with any other extension service the financial aspect is a key issue. Who pays for the services? ICT based extension services often involve private sector parties, like mobile phone service providers. An example is the Vodafone Farmer's club, a mobile price plan that offers special rates and a range of information services to farmers. Private sector actors will step out in case payment is not guaranteed (Aker, 2011). Profit oriented service are relatively more client oriented, as they depend on clients for income, and therefore more up-to-date and targeted. Customisation is a key requirement for sustainability

Key Lessons and Recommendations

The sheer multitude of ICT initiatives in agricultural extension plays an important role in revitalising the interaction between extension services and farmers by making services more demand-driven, up-to-date and inclusive. However, ICT are but one element in the wider transformation towards pluralistic extension services. Francis and Addom (2014) argue that extension alone cannot lift people out of poverty unless there is the right combination of policies, technologies, and market opportunities.

Context-specificity is critical. To be effective, it is essential to choose the ICT tool that is the most appropriate in a specific situation and context, such as in terms of affordability (McNamara *et al.*, 2011). This depends both on the type of information to be transferred as well as on the characteristics of the target group.

ICTs should not be seen as the sole solution to solving the challenges associated with agriculture, as broad access to more sophisticated and integrated ICTs requires organisational capacity. Public sector agricultural extension systems at present lack this capacity. The private sector could play a key role in changing current communication concepts in agricultural extension. However, this requires a supportive policy and enabling environment in order to facilitate the development of strong institutions and private sector actors who can drive innovations in ICT based extension.

Moreover, ICTs constitute supporting tools that can never fully substitute 'conventional' extension approaches like real-time advice from extension officers. In fact, without an organised extension system, ICT will be less likely to achieve widespread impact. In addition, ICT can neither do without nor replace face-to-face interaction between farmers and extension agents. Support to enhance extension systems in developing countries should therefore be broader than the introduction of ICTs.

Finally, empirical evidence on the impact of ICT remains limited. In order to measure the impact of such services on farmers' knowledge, adoption and welfare, as well as their cost-effectiveness, comprehensive impact evaluations are needed (Aker, 2011). In relation to impact evaluation, it is important to be aware of differential impacts and the factors that cause these differences, such as gender and social differences in access and use of ICTs (McNamara *et al.*, 2011).

As mobile phones and other ICTs continue to gain popularity and offer unique opportunities to share information among large numbers of farmers and other stakeholders, there is little doubt that ICT has a key role to play in agricultural extension systems. While not a panacea for all of the problems facing extension, the integration of ICTs as a communication channel potentially reaching millions of farmers and as an enabler of change is considered a critical part of future reform and investment agendas.

REFERENCES

Aker, J.C. 2011. Dial "A" for Agriculture: A Review of Information and Communication Technologies for Agricultural Extension in Developing Countries. *Agricultural Economics,* 42: 631–647.

Aker, J.C. and Mbiti, I. 2010. Mobile Phones and Economic Development in Africa. *Journal of Economic Perspectives,* 24(3): 207-232.

Bell, M. 2015. Powering Behaviour Change for a Brighter Agricultural Future. MEAS Discussion Paper, University of California, Davis.

Christoplos, I. 2010. Mobilizing the Potential of Rural and Agricultural Extension. Rome: Food and Agriculture Organisation of the United Nations (FAO).

FAO. 2011. The State of Food and Agriculture. Women in Agriculture: Closing the Gap for Development. Rome: FAO.

FAO. 2015. Success stories on information and communication technologies for rural development. RAP Publication 2015/02. Bangkok, Thailand: FAO Regional Office for Asia and the Pacific.

Francis, J. and Addom, B.J. 2014. Modern ICTs and Rural Extension: Have we reached the tipping point? Rural 21, 01/2014: 22-24. Grameen Foundation 2016. Community Knowledge Worker. [*Retrieved from at http://www.grameen foundation.org/what-wedo/agriculture/community-knowledge-worker*].

Kalas, P.P., and Finlay, A. 2009. Planting the knowledge seed adapting to climate change using ICTs: Concepts, current knowledge and innovative examples. Building Communication Opportunities (BCO) Alliance.

McNamara, K., Belden, C., Kelly, T., Pehu, E. and Donavan, K. 2011. Introduction: ICT in Agricultural Development. In: The World Bank (ed.), ICT in Agriculture Connecting smallholders to Knowledge, Networks, and Institutions (e-book). Washington, DC: World Bank.

Ministry of Information and Broadcasting. (2014). Facts and figures on community radio stations, Ministry of Information and Broadcasting, Government of India.

Nanda, S. and Arunachalam, S. 2009. Reaching the Unreached: Community based Village Knowledge Centres and Village Resource Centres, IGNOU Jamsetji Tata National Virtual Academy (NVA), MS Swaminathan Research Foundation, Taramani, Chennai.

Ospina, A.V., and Heeks, R. (2012). ICT-Enabled Responses to Climate Change in Rural Agricultural Communities. *Climate Change, Innova ion and ICTs Project-Strategy Brief*, 2, 5.

Press Release on Telecom Subscription Data as on 31 August, 2019"(PDF). *Telecom Regulatory Authority of India*. 18 October 2019 All Time Ranks (4 May 2019), *Most Subscribed Mobile Networks In India* (2009 -2019).

Quisumbing, A.R. and Pandolfelli, L. 2009. Promising Approaches to Address the Needs of Poor Female Farmers. Resources, Constraints and Interventions. IFPRI Discussion Paper 00882, July 2009. Washington: IFPRI.

Saravanan, R., Sulaiman, R., Davis, K. and Suchiradipta, B. 2015. Navigating ICTs for Extension and Advisory Services. GFRAS Good Practice Note for Extension and Advisory services.

TRAI. 2013. The Indian telecom services performance indicator, 2013

TRAI. 2014. Highlights on telecom subscription data as on 30th November, 2013.

TRAI. 2019. The Indian Telecom Services Performance Indicators"(PDF) (Press release). TRAI. 30 June 2019. p. 107-115.

TRAI-1. 2008. Measures to improve the telecom Penetration in Rural India, Next 100 million Subscribers, GOI, New Delhi.

USAID. 2010. ICT to Enhance Farm Extension Services in Africa. Briefing Paper. Available at http://pdf.usaid.gov/pdf_docs/PA00J7P8.

USAID. 2012. Gender Mainstreaming in ICT for Agriculture. Briefing Paper.

[*Retrieved form http://www.ngoconnect.net/documents/592341/749044/Gender+Mainstreaming+in+ICT+for+Ag*].

World Bank 2016. World Development Report 2016: Digital Dividends. Washington, DC: World Bank.

Chapter 7

Strategic Integrated Model to Achieve Viable Support for Entrepreneurship in Food Processing

U.S. Gautam[1], Vister Joshi[2] and Dheeraj Mishra[3]

[1]Vice Chancellor,
Banda University of Agriculture and Technology, Banda
[2]Subject Matter Specialist, Agriculture Extension,
Krishi Vigyan Kendra, Jalaun
[3]Assistant Professor, Agricultural Extension,
Banda University of Agriculture and Technology, Banda

Introduction

The food processing sector is critical to India's development, for it establishes a vital linkage and synergy between the two pillars of the economy-Industry and Agriculture. India holds the potential to acquire the numero uno status with sustained efforts in this sector. According to Ministry of Food Processing Industries, over the years agricultural production in India has consistently recorded higher output. India ranks third in the world in the production of Milk, Ghee, Pulses, Ginger, Banana, Guava, Papaya and Mango. Further, India ranks second in the world in the production of Rice, Wheat and many other fruits & vegetables. During the last 5 years ending 2018-19, Food Processing sector has been growing at an average annual growth rate of around 10 per cent. Food Processing is one of the major employment intensive segments. According to Ministry of Food Processing Industries, approximately 70.44 lakh people (19.33 Lakh registered and 51.11 lakh unregistered) are employed in food processing sector accounting 39,748 registered units (MoFPI, 2021). The industry provides 13 per cent to employment of manufacturing segment and contributes about 9.5 per cent to Gross Value Added (GVA). It is understood that this sector has enormous possibilities and more the entrepreneurship activities, better the development (MoFPI, 2021).

The term "entrepreneurship" comes from the French verb "entreprendre" and the German word "unternehmen", both mean to "undertake". Hofer and Bygrave (1992) defined the entrepreneurial process as 'involving all the functions, activities, and actions associated with perceiving of opportunities and creation of organizations to pursue them'. Entrepreneurs are the ones who explore opportunities, scan the environment, mobilize resources, convert ideas into viable business proposition and provide new products and services to the society by bringing together and combining various factors of production.

Entrepreneurship development and training is, thus, one of the key elements for development of micro and small enterprises (MSEs), particularly, the first generation entrepreneurs. To undertake this task on regular basis, the Ministry has set up three national-level Entrepreneurship Development Institutes (EDIs). These are the National Institute for Micro, Small and Medium Enterprises (NI-MSME), Hyderabad; the Indian Institute of Entrepreneurship (IIE), Guwahati and the National Institute for Entrepreneurship and Small Business Development (NIESBUD), Noida. Further, the Ministry has been implementing (in addition to the schemes of MSME-DO) an important scheme, namely, Scheme for Assistance for Strengthening of Training Infrastructure of Existing and New Entrepreneurship Development Institutes (EDIs).

In spite of all these efforts there are various problems faced by the entrepreneurs during incubation, maintaining and sustaining phases of entrepreneurship development. The problems faced by entrepreneurs involved in food processing unit have been at micro as well macro level, few problems like low capacity utilization, poor recovery of finished product from the raw materials, problems of arranging adequate working capital and its management, low product quality and unreliable assured power supply affects the entrepreneurship in this sector. Present chapter based on challenges envisaged by the entrepreneurs and the strategic action taken by different stakeholders to create a smooth entrepreneurial environment to reinvigorate the food processing sector.

Phases of Entrepreneurship Development

The entrepreneurship is a creative activity and an innovative function for running an enterprise. The process consists of three broad phases:

1. Stimulatory phase,
2. Support phase and
3. Sustaining phase

Stimulatory Phase

The initial phase of entrepreneurship development aims at creating awareness among people so that the entrepreneurship can be sprout. People are motivated

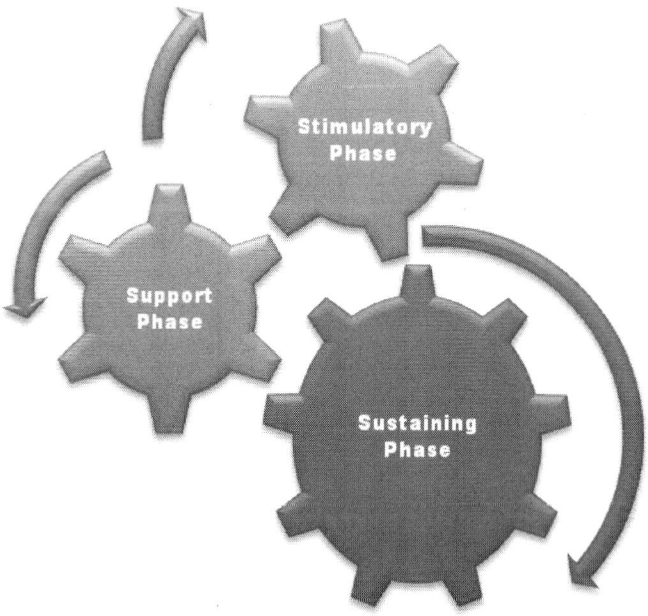

Figure 7.1: Phases of Entrepreneurial Cycle.

and offered with opportunities to acquire relevant information and skills. Key functions of this phase are:

1. Generating entrepreneur awareness in the community through well planned publicity.
2. Identifying and potential entrepreneurs.
3. Developing managerial, technical, financial, and marketing skill

Support Phase

The phase starts from incubation of idea and converting it into a viable venture. All aspects for smooth running of enterprise come into support phase. Thus, hindrances against the smooth running of units at the time of genesis, management, institutional, financial, technological and marketing are addressed during this phase (Figure 7.2).

Sustaining Phase

The phase refers to the activities facilitating growth and continuity of entrepreneurial venture through expansion, modernization, diversification and technology up gradation of on-going enterprises and rehabilitation of sick units. The following activities are undertaken during sustaining phase:

Figure 7.2: Functions Support Phase.

1. Modernization and product substitution
2. Additional financing
3. Expansion and diversification of business activities
4. Diagnosing and reconsidering the factors causing failure

Problems at Different Phase of Food Processing Enterprise Development Encountered By Entrepreneurs'

Stimulatory Phase	*Sustaining Phase*
☆ Lack of strong economic environment to establish a food processing unit ☆ Government taxes and policies ☆ Choice and selection of product ☆ Lack of technological environment to encourage entrepreneurship in food processing sector ☆ Lack of entrepreneurial awareness in the communities ☆ Lack of political will to develop food processing enterprises ☆ Lack of proper training to incubate the entrepreneurial skill and knowledge	☆ Lack of additional financing for full capacity utilization of the unit ☆ Deferring repayment or interest depending on situation ☆ Lack of guidance in diagnosing the failure or low production or profit ☆ Food processing is highly fragmented and unorganized sector ☆ Lack of infrastructure (mainly poor supply chain) ☆ Quality concern

Support Phase	
Genesis ☆ Licensing and registration of unit ☆ Lack of techno-economic information ☆ High cost for preparation and development of project ☆ Lack of data bank on new products and process available to the target group ☆ Purchase of machinery and equipments	**Financial** ☆ High rate of interest ☆ Inadequate amount from financing agencies ☆ Financial management ☆ Difficulty in recovering from the buyers ☆ Lack of sufficient fund
Management ☆ Lack of skilled labour ☆ High wages of labours ☆ Low scale of production ☆ Lack of trained staff ☆ Poor communication between different departments	**Technical** ☆ Lack of technical know how ☆ Fast advancement in technologies ☆ Lack of maintenance of machineries and equipment's ☆ Non availability of advanced technologies in time ☆ Outdated technology
Institutional ☆ Lack of proper linkage with the different institutions ☆ Improper extension system of research institutes ☆ Lack of R&D institutes to impart need based local technologies ☆ Lack of Government support and incentives ☆ Long and complicated procedure to avail institutional help	**Marketing** ☆ Competition from larger units ☆ Superior and cheaper competitive substitute ☆ High cost of packaging ☆ Unfamiliarity with export activities ☆ High import and export duties on packaging material

Under the Atmanirbhar Bharat mission, this is the time to transform the food processing industry into demand driven sector. Appropriate measures need to be taken for reducing costs, ensuring quality production, improving market access, creating efficient supply chain equipped with advanced technological interventions and management (Singh *et al.*, 2021).

Strategic Planning and Support to Entrepreneurs engaged in Food Processing

The entrepreneurs in food processing sector are encountered with numerous problems during enterprise development. The conceptual model presented in Figure 7.3 suggests the remedial measures to cope up with the problems. The figure shows that to support food processing entrepreneurship, there must be a synergy between Government's Action (GA) and Institutional Actions (IA), through public private partnership and it should be assisted at first stage mentioned as "Preparatory phase". This indicates that government assistance in terms of realizing the support at incubation of food processing units. This could be achieved with individual as well as integrated efforts of each component. These are described here as,

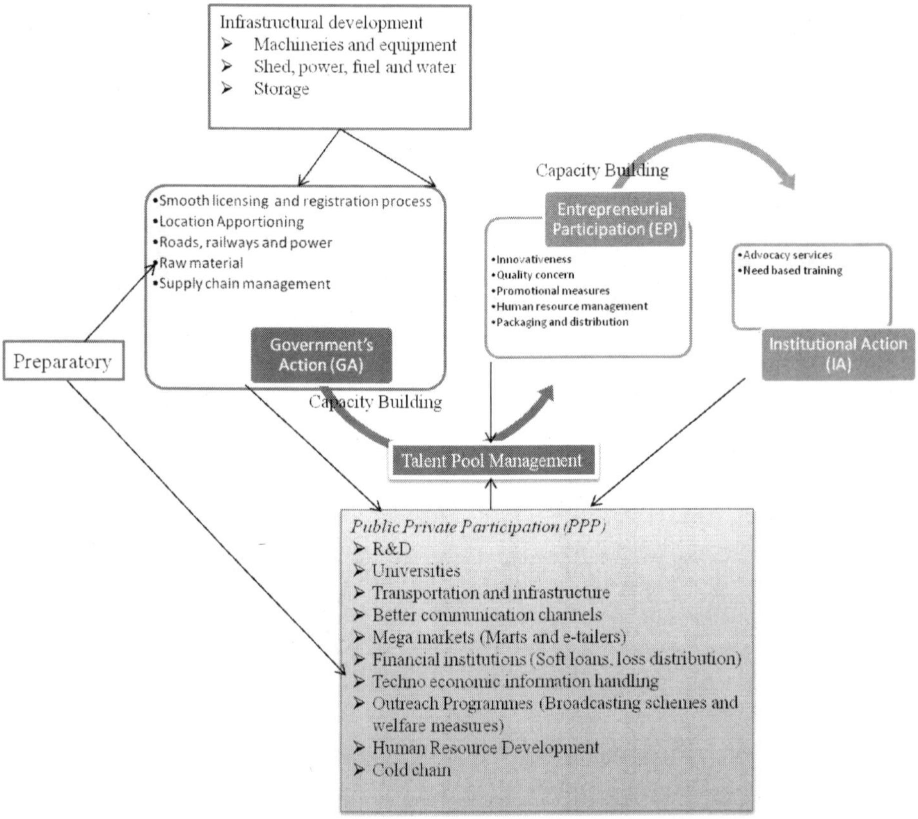

Figure 7.3: Strategic Planning and Support (SP&S) Model to Strengthen Entrepreneurship.

Streamlining Government Action (GA)

Government must streamline its action and develop total factors in state to support food processing units. Government must smooth the licensing and registration process to encourage incubation of new units into the system. "Location Apportioning" is another considerable issue as majority of registered entrepreneurs was not satisfied with the location of unit. Thus, government must focus on providing land at subsidized rate particular, for the enterprise like food processing. To achieve total factor productivity, government must focus on developmental issues. Proper transportation system (Roads and Railways) and power should be developed up to rural areas of state. Raw materials were considered as another important area of problem as timely availability in ample quantity could not achieved. Thus, government must ensure warehousing and storage system for procurement of raw material.

Streamlining Institutional Action (IA)

Institutions should be linked properly with government either as a government agency or as public private participation. The model clearly differentiates the roles of different component in a specific manner. Capacity building is one of the major role of institutions which should be strengthened through strong and efficient extension system (outreach), which is possible with government interventions with wider space for private agencies. Secondly, infrastructure in terms of Transportations, communication, financial institutions, and supply chain should be well streamlined through Public private partnership only. Adequate and timely financial assistance is prerequisite for a viable unit. After repeated recommendation by planners and researchers, the financial institutions are still not in justice with their responsibilities for providing sufficient fund.

Entrepreneurial Participation (EP)

It must ensure by the government and institutions that growth could achieve only with the participation of entrepreneurs. Entrepreneur has a critical role and they must be proactive in adoption and innovation. Another important suggestion is regarding quality control as with government, it is also a major responsibility of entrepreneur to ensure quality product in market. Thus, Government-entrepreneur-institutions ensures not only quality but ethical competition in the market which also maintain prices of products and give equal chance to SME as well as large enterprises to sustain in market.

Thus, to achieve sustainability in entrepreneurship in food processing industries, support phase should be strengthen with considering total factors involving in its development.

Conclusion

The strengthening of the food processing sector requires entrepreneurship and innovation. This chapter provides an overview of entrepreneurship development in light of food processing sector and problem associated with the entry as well sustainability of these units at micro as well macro business environment. Further, based on the issues envisaged by the food processor at different point of time, an innovative model at planning and support phase of entrepreneurship development presented. The model (SP&S) emphasized on streamlining the actions taken by the one stakeholder in linked manner thus to mutual support the action of other stakeholders engaged in food processing. This chapter will further open the point of discussion for policy makers, researchers and students engaged in entrepreneurship studies in different sects to render their services in better ways to entrepreneurs concerns. The presented concept can use to create value for food processing entrepreneurship also, over time, contribute to changes in practices in the industry.

REFERENCES

Hofer, C.W. and Bygrave, W.D. 1992. Researching Entrepreneurship. Entrepreneurship Theory and Practice, 16(3): 91-100.

Singh, G., Daultani, Y. and Sahu, R. 2021. Investigating the barriers to growth in the Indian food processing sector. *OPSEARCH*. https://doi.org/10.1007/s12597-021-00553-1

https://www.investindia.gov.in/sector/food-processing

https://mofpi.nic.in/documents/reports/annual-report

Chapter 8

Modern ICTs Tools in Agricultural Extension for the Empowerment of Smallholder Farmers: Cases from Bihar Agricultural University, Sabour, India

R.K. Sohane[1] and Aditya Sinha[2]

[1]Director, Extension Education and [2]Assistant Professor, Department of Extension Education, Bihar Agricultural University, Sabour, Bihar

Introduction

The start of the 21st Century poses emerging challenges in relation to global food demand and exceeding the earth's carrying capacity with the current way of agricultural production. Virtual technologies backed by Information and communication technologies (ICTs) have played an important role in agriculture and can be said as well timed in relation to current challenges faced in agriculture and allied disciplines. The role of green revolution in improving agricultural productivity can never be undermined but the need of a new technology leading to better marketing linkages, efficient advisory system and reduced wastages in agricultural systems has long been felt. ICT is no more a fancy deal. It has taken a giant step forward surpassing costly, bulky and energy-consuming tools to those which is cost-effective, easily accessible and energy saving. The exponential growth of mobile phones, wireless and internet services has led to better private sector participation in agricultural research and development along with the application of latest tools for agricultural growth. Mobile phone penetration in the developing world now exceeds two subscriptions for every three people which are still expanding in Asia and Africa (Abdulsalam *et al.*, 2016). The technologies associated with artificial intelligence and virtual systems in agriculture like the GPS based communication satellites, enabling of wireless transmissions, to the traditional technologies like the radio and the television (Sharma, J.P. *et al.*, 2016). The virtual capabilities are facilitating the development, sharing, storage, retrieval and dissemination of a range of information and support communication in many forms (*e.g.* Email, discussion lists, chat forums, video). These are proving to be quite helpful in conducting extension on a wider spectrum with increased penetration.

The high data speed has made possible the use of cloud computing services for managing and delivering extension services rapidly. The future calls for technologies which could combine drone technology, wireless sensor networking, Internet of Things (IoT), Big Data and geospatial information systems to offer new solutions to Indian agriculture (Gill, S. S. *et al.*, 2017). The flagship scheme of the Government of India in promoting the use and development of internet-based programmes for the farming community called 'Digital India' is providing for a powerful infrastructure and service platform which could be instrumental in providing lot of solutions to the farming community with enhanced speed. Also, the rapid proliferation of custom hiring centers is leading to better mechanization of the agricultural farm holdings (Ankur, S. and Ganguly, K., 2017).

Table 8.1: Data of Telecom Subscription in India

Particulars	*Wireless*	*Wireline*	*Total (Wireless + Wireline)*
Total Telephone Subscribers (million)	1156.44	20.58	1177.02
Urban telephone Subscribers (million)	644.54	18.21	662.75
Rural telephone Subscribers (million)	511.90	2.37	514.27
Overall tele-density (per cent)	85.92	1.53	87.45
Urban tele-density (per cent)	140.20	3.96	144.16
Rural tele-density (per cent)	57.76	0.27	58.03
Share of Urban Subscribers	55.73 per cent	88.47 per cent	56.31 per cent
Share of Rural Subscribers	44.27 per cent	11.53 per cent	43.69 per cent

Source: TRAI (2020). Highlights of Telecom Subscription as on 8 May 2020; data available on https://www.trai.gov.in/sites/default/files/PR_No.29of2020.pdf

The data on telecom subscription in India for the year 2020 is discussed in Table 8.1. It can be observed from the table that the share of telecom subscribers in urban area is 140.20 per cent compared to 57.73 per cent in the rural areas. Thus, the subscription to mobile services has reached a saturation in the urban areas while the rural areas has just crossed the half-way mark. The monthly growth rates of urban and rural subscription are nearing or overl per cent in the current decade (TRAI, 2020). The main contributing factors responsible for the proliferation of ICT in agriculture are low-cost and pervasive connectivity, adaptable and more affordable tools, advances in data storage and exchange along with innovative business models and partnerships. The democratization of information including the open access movement and rise of social media to project oneself are major drivers for shaping the prospects for using ICT effectively in developing countries. The various ICT in agriculture initiatives of the university is discussed in detail below:

Digital Storytelling though Online Digital Videos

Digital storytelling is a short form of digital media production that allows everyday people to share aspects of their story with the use of video as a medium for increased adoption and behavioural change. It has several possible advantages for low resource settings, including low cost (Cox RH, White AH, Gaylord CK, 2003) accessibility by illiterate or semi-literate populations, standardization of the information communicated (Gagliano M., 1988) and easy dissemination of information to the intended beneficiaries. The interest of farmers in watching a particular video depends strongly on its content. The videos demanded by the farmers are basically based on multiple farmers adopting the same practice (Gandhi, R. *et al.*, 2009). Electronic Media and Production Centre (EMPC)was established in 2011 at the university for the production of audio, video and multimedia means of interacting with farming community. The center has developed technical videos on more than 110 subjects like Integrated Farming Systems, Basil cultivation, Betelvine cultivation, Dragon fruit farming, Quail farming, Poultry farming, Scientific Method for making Tomato Ketchup, Goat farming, Honeybee rearing (*Madhumakhi palan*), cultivation of pointed gourd, strawberry cultivation in Bihar and others. The videos are uploaded on YouTube and are receiving continued interest of the farming community. The total number of subscribers to the YouTube channel has crossed the 263 thousand mark while the total viewership is above 26 million. The Electronic

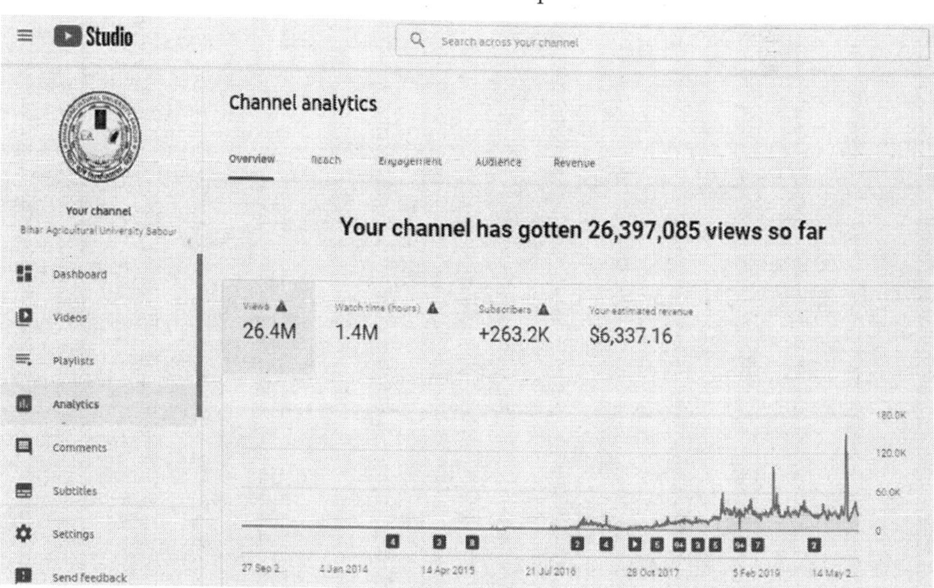

Figure 8.1: Snapshot Showing Youtube Overview of BAU Sabour Channel as on 16 May 2020.

Media Production Center at the university is also supporting Doordarshan for agricultural knowledge dissemination.

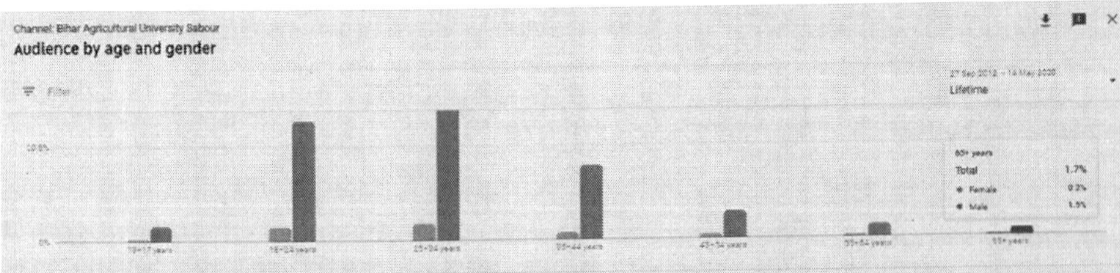

Viewer age	Views Female	Views Male	Watch time (hours) Female	Watch time (hours) Male
Total	**9.4%**	**90.6%**	**8.9%**	**91.1%**
13–17 years	0.4%	3.0%	0.3%	2.5%
18–24 years	2.9%	25.3%	2.4%	22.3%
25–34 years	3.5%	37.0%	3.2%	36.6%
35–44 years	1.4%	15.6%	1.4%	17.1%
45–54 years	0.8%	5.6%	0.9%	7.0%
55–64 years	0.4%	2.6%	0.5%	3.6%
65+ years	0.2%	1.5%	0.2%	1.9%

Figure 8.2: Distribution of Viewership Based on Age Group (16 May 2020).

The Figure 8.2 above shows that the videos are quite popular among the youth in the age group of 18-24 and 25-34 years with sustained watch time. This clearly shows that the videos are of interest to the youth who may be using it in the development of projects on agriculture.

The Figure 8.4 shows that the video on Layer poultry farming have gathered maximum attention among the viewers with viewership of over 33 lakh people followed by success story of backyard poultry farming with 17.02 lakh viewership, quail farming with a viewership of 16.09 lakh, pineapple cultivation in Bihar with 12.68 lakh and cultivation of pointed gourd with 10.29 lakh viewers.

Figure 8.3 shows that the videos watched is maximum from India with 87.30 per cent watch time followed by Pakistan with 2.50 per cent, Saudi Arabia (2.20 per cent), Nepal (1.30 per cent) and United Arab Emirates (1.20 per cent). The videos

Geography	Views		Average view duration	Watch time (hours)	
Total	**26,397,080**		**3:08**	**1,383,081.4**	
India	23,099,453	87.5%	3:08	1,206,911.7	87.3%
Pakistan	653,772	2.5%	2:45	29,990.2	2.2%
Saudi Arabia	582,358	2.2%	3:28	33,782.6	2.4%
Nepal	353,987	1.3%	3:04	18,124.3	1.3%
United Arab Emirates	317,290	1.2%	3:16	17,333.1	1.3%
United States	147,551	0.6%	4:23	10,797.1	0.8%
Kuwait	143,151	0.5%	3:43	8,897.7	0.6%
Bangladesh	115,038	0.4%	2:13	4,251.6	0.3%
Qatar	109,767	0.4%	3:25	6,263.0	0.5%
United Kingdom	80,981	0.3%	3:59	5,392.3	0.4%

Figure 8.3: Demographics of Videos Watched from Various Countries (as on 10 May 2020).

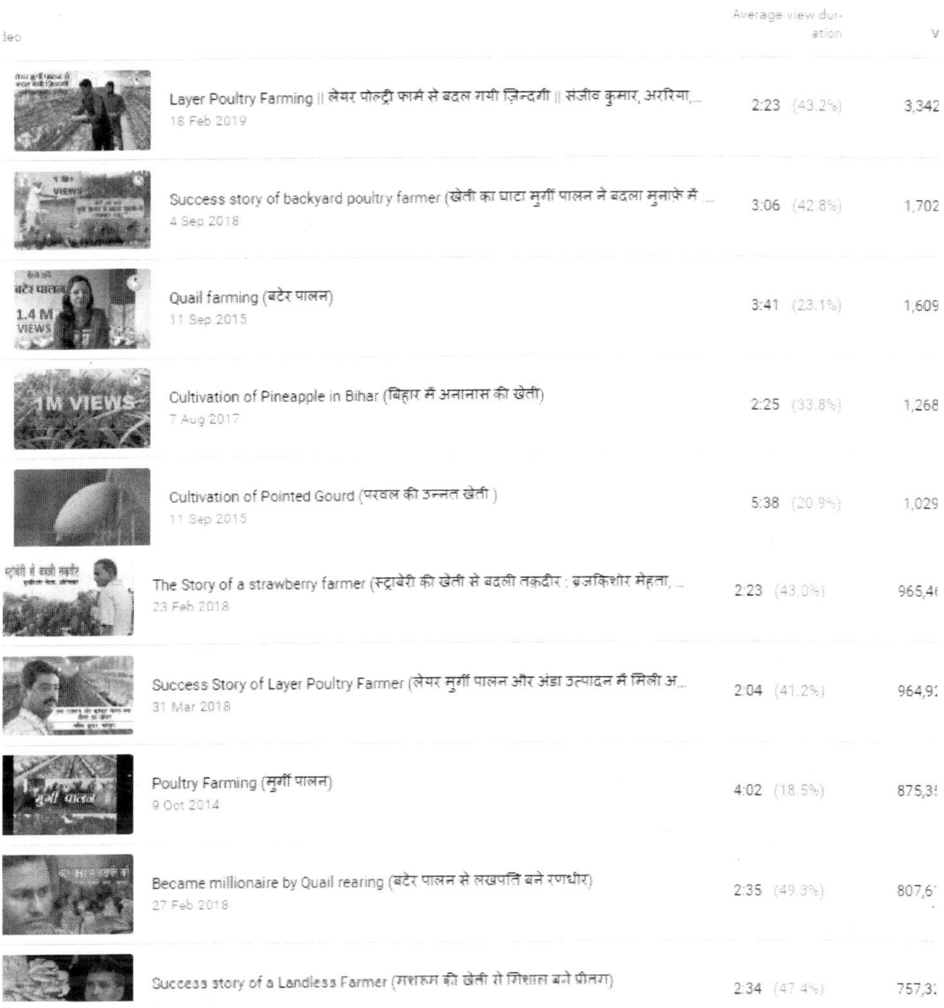

ieo	Average view duration	V
Layer Poultry Farming ॥ लेयर पोल्ट्री फार्म से बदल गयी ज़िन्दगी ॥ संजीव कुमार, अररिया,... 18 Feb 2019	2:23 (43.2%)	3,342
Success story of backyard poultry farmer (खेती का घाटा मुर्गी पालन ने बदला मुनाफे में ... 4 Sep 2018	3:06 (42.8%)	1,702
Quail farming (बटेर पालन) 11 Sep 2015	3:41 (23.1%)	1,609
Cultivation of Pineapple in Bihar (बिहार में अनानास की खेती) 7 Aug 2017	2:25 (33.8%)	1,268
Cultivation of Pointed Gourd (परवल की उन्नत खेती) 11 Sep 2015	5:38 (20.9%)	1,029
The Story of a strawberry farmer (स्ट्राबेरी की खेती से बदली तक़दीर : ब्रजकिशोर मेहता, ... 23 Feb 2018	2:23 (43.0%)	965,4
Success Story of Layer Poultry Farmer (लेयर मुर्गी पालन और अंडा उत्पादन में मिली अ... 31 Mar 2018	2:04 (41.2%)	964,9
Poultry Farming (मुर्गी पालन) 9 Oct 2014	4:02 (18.5%)	875,3
Became millionaire by Quail rearing (बटेर पालन से लखपति बने रणधीर) 27 Feb 2018	2:35 (49.3%)	807,6
Success story of a Landless Farmer (मशरूम की खेती से मिशाल बने प्रीतम)	2:34 (47.4%)	757,3

Figure 8.4: Web Analytics of Farmer Friendly Videos Shared on YouTube (as on 10 October 2019).

have generated lot of interest in improved agricultural practices with regular comments from the viewers.

Providing Better Connectivity and Advisories through Video Conferencing

Bihar Agricultural University, Sabour has moved a step further in providing connectivity to all the twenty KVKs via high-bandwidth video conferencing facilities namely Araria, Arwal, Aurangabad, Banka, Bhagalpur, Gaya, Jehanabad, Katihar, Khagaria, Kishanganj, Lakhisarai, Madhepura, Munger, Nalanda, Patna, Purnea, Rohtas, Saharsa, Sheikhpura and Supaul are connected through headquarter. It provides an opportunity for two-way interaction among the scientists, extension workers and farming community quickly and efficiently. The direct two-way communication of the farmers with experts help in better understanding of the farmers' problems and in providing the timely and instantaneous solutions. Video conferencing facility has helped in increasing the productivity along with the improving the communication and strengthening of farmer-scientist linkage. It resulted in cutting down of travel expenses and saving of time associated with conventional methods of interaction.

Table 8.2 shows that there is an increased organization of training programmes with the technical assistance of scientist from the university headquarters. The number of video conferencing organized has increased from 472 in the year 2017-18 to 520 in the year 2018-19 while 417 video conferencing were conducted in the year 2019-2020.

Adding Further Services through Live Video Solutions

The live video solutions enabled through TVU systems which are lightweight, portable video transmission systems capable of transmitting information from the farmers' field is also implemented in the university through which it is possible to link farers' fields virtually with the scientist sitting at headquarters for providing advisory to farmers on real-time basis. The impact of the video-conferencing and TVU systems is that it has led to increased efficiency of extension system by reaching large number of farmers at lower cost, quickly and simultaneously along with direct interaction between farmers and university scientists has strengthened the Farmer-Scientist- Extension worker linkage. The knowledge sharing on various aspects based on needs of the farmers are organized regularly with the active collaboration of the KVKs. The participation of farmers is on the rise with this innovative method as it saves time, money and effort. The domains covered under advisory through video conferencing are weed management, vaccination in poultry, orchard management, goatery, nutrient management, mushroom farming *etc.*

Provision of Farm Advisories through Kisan Call Center and Whatsapp Mobile Application

The university is reaching the farmers on telephone and whatsapp mobile application. The Kisan call center offers solution to the farmers on all aspects of agriculture on a dedicated toll-free number. In order to get specialized and customized advisories, the farmers also send pictures of the disease, problems on the farm through whatsapp mobile application for better advisories.

Providing Agricultural Information through a Dedicated Community Radio

The community radio is an important tool to empower rural community through quality information concerning their locality. The community radio station was established on May 17, 2011 at KVK, Barh, Patna on 91.2 F.M. radio band with the specific objectives to broadcasts content that is relevant to a local, specific audience but is often overlooked by commercial or mass-media broadcasters. The

Table 8.2: Training Organized and Participation of Farmers from Various KVKs through Video Conferencing

Sl. No.	Name of KVK	No. of Video Conferencing			No. of Farmers Benefitted								
		2017-18	2018-19	2019-20	2017-18			2018-19			2019-20		
					Male	Female	Total	Male	Female	Total	Male	Female	Total
1	Araria	38	26	20	1778	412	2190	1012	256	1268	955	128	1083
2	Arwal	19	23	20	819	169	988	997	210	1207	1012	140	1152
3	Aurangabad	28	25	22	1243	396	1639	1003	315	1318	980	270	1250
4	Banka	26	27	24	1156	248	1404	1042	281	1323	822	218	1040
5	Bhagalpur	38	31	23	1645	481	2126	1118	401	1519	945	281	1186
6	Gaya	19	27	23	870	156	1026	1071	256	1327	1001	256	1257
7	Jehanabad	27	26	27	1114	344	1458	933	341	1274	1014	144	1158
8	Katihar	30	29	22	1242	378	1620	1096	325	1421	900	570	1470
9	Khagaria	15	24	16	693	117	810	971	205	1176	7887	1005	
10	Kishanganj	18	27	21	878	112	990	1066	269	1335	944	112	1056
11	Lakhisarai	16	26	18	862	110	972	1038	218	1256	890	119	1009
12	Madhepura	16	25	14	608	112	720	1025	206	1231	508	122	628
13	Munger	19	25	18	756	270	1026	979	257	1236	855	275	1130
14	Nalanda	19	28	20	800	245	1045	1007	265	1372	850	256	1106
15	Patna	18	28	19	794	180	974	1080	301	1381	780	215	995
16	Purnea	18	26	27	818	162	980	1004	254	1258	908	162	1070
17	Rohtas	34	27	25	1231	437	1668	1097	239	1336	880	141	1021
18	Saharsa	36	24	19	1452	463	1915	962	225	1187	580	125	605
19	Sheikhpura	19	21	20	944	101	1045	800	229	1029	809	160	969
20	Supaul	19	25	19	937	95	1032	949	296	1245	780	104	884
	Total	**472**	**520**	**417**	**20640**	**4988**	**25628**	**20250**	**5349**	**25699**	**17201**	**4015**	**21074**

area of Community Radio Station is 8 Blocks, 72 Village Panchayat and 503 Villages of Patna District. The Community Radio is a model of Radio broadcasting in addition to commercial and public broadcasting offered by the extension departments and the directorate. Community Radio Station has a great potential for education and social development of farming communities. It is most economical tool to reach the masses and have a quick communication of technologies and information related to agriculture and allied sector.

Table 8.3: Programme Broadcasted on Community Radio Station during 2018-19

Sl.No.	*Morning Hours*			
	Programme	*Duration of Programme*	*No of Days/Year*	*Total Hrs.*
1.	Krishak Manch	30 Minutes	278	139
2.	Success Story of farmers	30 Minutes	278	139
3.	Swasth Churcha	15 Minutes	278	69.5
4.	Lok Rang	15 Minutes	278	69.5
	Total	**1.30 Hrs.**	**278**	**417**
Sl.No.	*Evening Hours*			
	Programme	*Duration of Programme*	*No of Days/Year*	*Total Hrs.*
1.	Krishak Manch	45 Minutes	282	211.5
2.	Mahila Jagat	15 Minutes	282	70.5
3.	Bal Manch	15 Minutes	282	70.5
4.	Lok Rang	15 Minutes	282	70.5
	Total	**1.30 Hrs.**	**197**	**423**

Table 8.4: Programme Broadcasted on Community Radio Station 90.8 FM Green during 2019-20 (5 Aug 2019 to 31 March 2020)

Sl.No.	*Programme*	*Broadcast*	*Duration of Programme*	*No of Days/Year*	*Total Hrs.*
1.	Unnat Krishi	Daily	30 Minutes	240	120
2.	Swasth Rahen Hum	Daily	15 Minutes	240	60
3.	Carrier Watch	Alternate	15 Minutes	120	30
4.	Ang Umang	Daily	30 Minutes	240	120
5.	Krishi Ki Kahawten	Alternate	15 Minutes	120	30
6.	Mahila Jagat	Daily	30 Minutes	240	120
7.	Hamare Safal Kisan	Daily	30 Minutes	240	120
8.	Man Ki Baat (Live)	As per PMO information	40 Minutes (average)	10	6.6
Total				**240**	**756.6**

The community radio is broadcasting programmes like panel discussion on agriculture, success stories of farmers, special programme for women and children, health related soaps, and folk songs. The community radio is covering more than 13 lakhs farmers, farm women and children of the area under its jurisdiction.

From the last year, the community radio station namely *FM Green* has also started functioning in the university headquarters at Sabour 5th August 2019 to benefit the farmers of the region on 90.80 MHz frequency. It broadcasts programmes every day between 2-5 pm in Hindi and Angika language.

Other ICT Related Initiatives for the Farming Community

Providing Advisories on the Go through Kisan Gyan Rath

It is an important mobile system for imparting knowledge to the farmers at their doorsteps. *Kisan Gyan Rath, which* is equipped with facilities for soil sample analysis, agro-exhibits, extension literature and technical videos developed by BAU, Sabour. This mobile technology package has enabled the university to deliver its technological knowledge base and diagnostic services at the doorstep of farmers. This *Rath*moves from village-to-village as per the schedule across the 25 districts of Bihar. It is paying a pivotal role in motivating the farmers to adopt the *latest technologies.* Kisan Gyan Rath was launched by Bihar Agricultural University is *Kisan Gyan Rath*was launched by the Hon'ble Chief Minister of Bihar on 11 October, 2014. It is an important medium to provide expert guidance to the farmers on their doorsteps. The *Kisan Gyan Rath* (*Bus*) is a fully equipped having facilities for soil analysis, interpretation and issuing of soil health cards to the farmers. The benefit from the scheme is that farmers need not to visit the soil testing laboratories instead the soil testing laboratories reach farmers on their doorsteps. In 2017-18 the Kisan Gyan Rath has covered 426 villages of 63 blocks of 12 districts benefitting over1.25 lakh farmers directly.

Massive Open Online Course on Entrepreneurship Development in Agriculture

The university offered in course on entrepreneurship development in agriculture for employment generation among the rural youth in collaboration with IIT, Kanpur and Commonwealth of Learning during the year 2019-20. The course was an attempt to reach the rural youth and promote them to come forward as agri-entrepreneurs (Sohane, R.K. and Sinha, A., 2019). Some of the modules covered under the course were the principles of entrepreneurship, self-appraisal and management, various opportunities in the agricultural sector and development of a successful enterprise *etc.*

Way Forward

The integration of various ICT platforms like YouTube, WhatsApp, Videoconferencing as well as offline services like SD Cards, Community radio stations as well as mobile messaging services is effectively implemented by the University. In recognition of the use of ICT for technology transfer, the National e-Governance award-2020 and Technology Sabha Award for its outstanding contribution in the area of employing ICTs for agricultural development in the state of Bihar has been awarded by Indian Express Group. The YouTube channel of the university also received the "YouTube Creator award" by YouTube (Google Inc.) for production of farmer friendly technical videos and reaching more than one million potential farmers. This model has been considered for upscaling by Bihar State Govt. under its Krishi Road Map, 2017-22.

Conclusion

The paper concludes that there is an immense potential to reach the farming community effectively, generate interest in agriculture and promote entrepreneurial mindset through the effective implementation of ICT-based tools in agricultural extension. It is a vital machinery to support both the extension functionaries and the farmers. The timely reach of information through it bridges the barrier between the developers of technology and the practitioners leading to an enhanced interest. The low-cost mobile phones and internet connectively provides a way forward to multiply the initiatives in a phased manner with ease. Such innovative ICT-based projects need to be scaled up so as to make effective generalizations along with promoting quicker replication potential.

REFERENCES

Abdulsalam, A. Olaifa, T.P. and Frederick, A. 2016. The Complimentary Role of Information Communication Technology (ICT) in Agricultural Knowledge Management in Nigeria. *Journal of Agricultural Sciences*, 6(5): 173-179.

Cox, R.H., White A.H. and Gaylord, C.K. 2003. A video lesson series is effective in changing the dietary intakes and food-related behaviours of low-income homemakers. *J Am Diet Assoc.*, 103: 1488–1493. doi: 10.1016/S0002-8223(03)01230-6. pmid:14576714

Gagliano, M. 1988. A literature review on the efficacy of video in patient education. *J Med Educ.*, 63: 785–792. pmid:3050102

Gandhi, R., Veeraghavan, R., Toyama, K. and Ramprasad, V. 2009. Digital Green: Participatory Video and Mediated Instruction for Agricultural Extension. *Information Technologies and International Development*, 5(1): 1-15.

Gill, S.S., Chana, I., and Buyya, R. 2017. IoT based agriculture as a cloud and big data service: the beginning of digital India. *Journal of Organizational and End User Computing (JOEUC)*, 29(4): 1-23.

Seth, A. and Ganguly, K. 2017. Digital technologies transforming Indian agriculture. *The Global Innovation Index*, 105-111.

Sharma, J.P., Burman, R.R. and Gill, R. 2016. Overreaching farming through virtual extension techniques. *Indian Farming*, 66(8): 38-41.

Sohane, R.K. and Sinha, A. 2019. Modern ICTs tools in Agricultural Extension for the Empowerment of Small holder farmers: Cases from Bihar Agricultural University, Sabour, India. In: Souvenir cum Lead/Abstract Proceedings Book: Indian Society of Extension Education, pp. 66 – 72.

http://www.cto.int/wp-content/themes/solid/_layout/dc/kr/ICT_in_AgSourcebookweb light.pdf

http://www.iwss.info/downloads/files/n50d33368b217e.pdf

Chapter 9

Paradigm Shift in Role of Extension Educationists in the Light of e-Mediated Extension Services

Basavaprabhu Jirli

Professor and Head, Department of Extension Education, Institute of Agricultural Sciences, Banaras Hindu University, Varanasi, Uttar Pradesh

Introduction

Enough literature is available to understand the concept of extension. But are there any efforts to understand the in-tensions of ex-tension? Why should we emphasize on inner tensions of extension? To be a successful extension professional for bringing about the desirable changes in the behavior of stakeholders, we need to understand the inner-tensions of extension in addition to the concept of extension.

Since inception of extension services, the model adopted for transfer of technology was dominated Technology, wherein the core emphasis was on the transfer of technology. The end product was the partial adoption of the technology or non-adoption of the technology. Learning from the experiences, modifications were made in the model and came out with extension mediated technology transfer system, wherein the technology is inseparable component of extension, which was combined with the human element or the societal element. The approaches adopted by different models of extension service delivery were also diverse.

The first model of extension service delivery the Public extension system or the Ministry of Agriculture, which delivers the extension services in the form of a systems approach. The relationship with the farmers or the end users is 'take it or

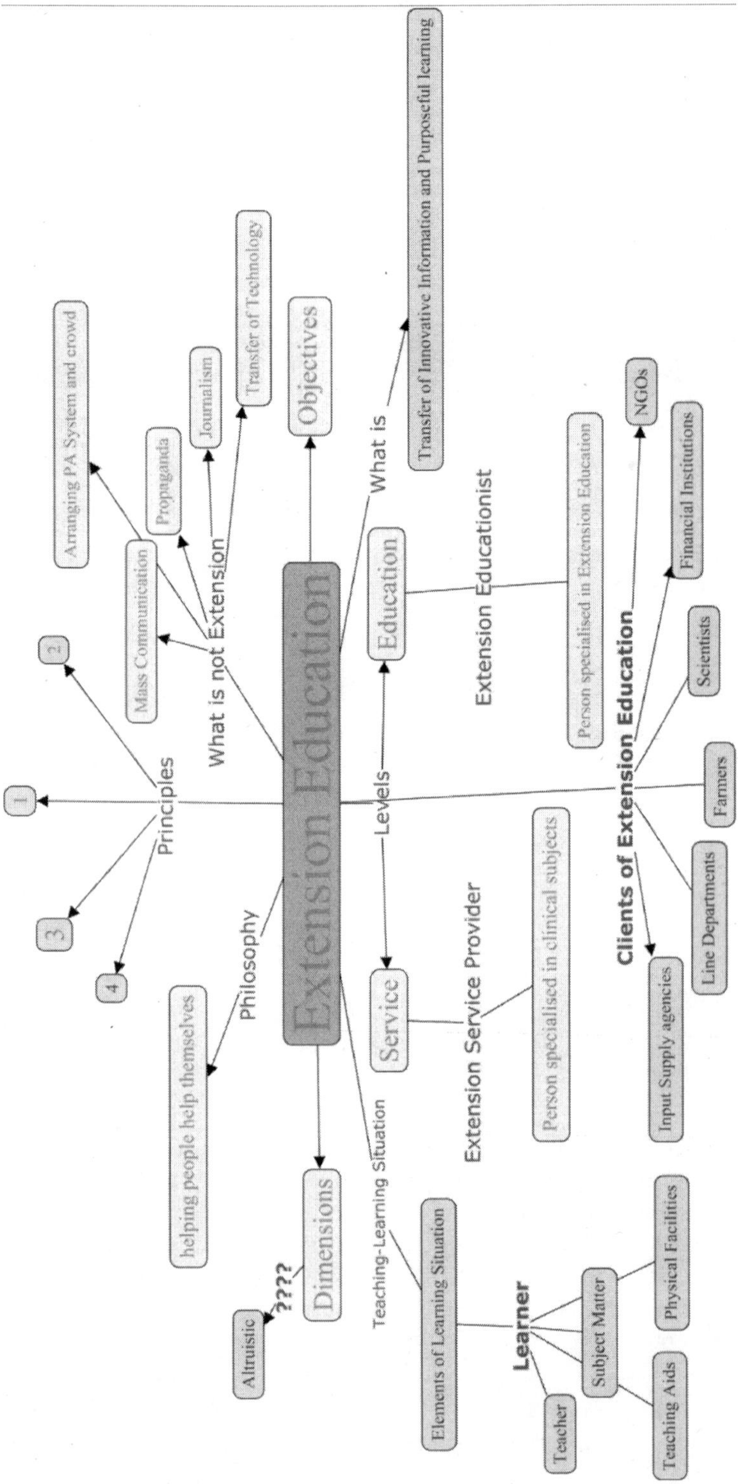
Extension Education
What is not Extension
Mass Communication
Arranging PA System and crowd
Propaganda
Journalism
Transfer of Technology
Objectives
What is
Transfer of Innovative Information and Purposeful learning
Education
Extension Educationist
Person specialised in Extension Education
Clients of Extension Education
NGOs
Financial Institutions
Scientists
Farmers
Line Departments
Input Supply agencies
Levels
Service
Extension Service Provider
Person specialised in clinical subjects
Principles
1
2
3
4
Philosophy
helping people help themselves
Dimensions
????
Altruisic
Teaching-Learning Situation
Elements of Learning Situation
Learner
Teacher
Subject Matter
Teaching Aids
Physical Facilities

leave it'. Whether the receivers are gaining it or not, whether they are understanding it or not, ultimately we are transferring it, and are ending it with the target oriented approach.

The second model was participatory approach of extension service delivery. The relationship with the farmers was 'take it or demand different packages of programs', wherein we can see the approaches of extension that were adopted in the post 1990's have invariably participatory in nature. People started demanding the things, wherein the Ministry of Agriculture and farmers associations came together.

The third model was the private extension efforts, the examples like the contract farming, wherein everything is compulsory. This model emphasizes on profit. The approach and the relationship with the end users is more of a 'compulsion on the receivers'. Many reports till date reveal that, the end users are being exploited to the maximum extent because of the lack of the legal support systems.

In all these the three models the end user is the most deprived entity. How we can overcome these type of situations? The way out can be by understanding the basic processes of extension, the fundamentals of extension, which include in the definition of the In-Tension of Ex-Tension. **The concept of In-Tension of Ex-Tension has been conceptualized as the degree to which the philosophy, the content, ideas, objectives, techniques, principles, theory and models of extension have been internalized by extensionist.**

Let's understand the concept of In-Tension of Ex-Tension in right perspective. As an extension professional while delivering the extension services, the first and foremost question to be asked to oneself is, whether I am making use of principles of extension in my service approach?, Whether I am delivering extension services as per objectives of extension? Whether I am exploiting the dimensions of extension in my service delivery efforts? Whether can I visualize implications of the processes of extension and methods of extension? These are the basic questions that extension professionals need to put to oneself before delivering extension services or acting as an extension professional.

The moment we ask such these questions, the emerging answer is leading towards the holistic approach of extension. That is what is the role of an extension educationist, to give due emphasis on the inner-tensions of extension. The extension educationist is providing you the basic inputs that are necessary for the process of transfer of technology. When we go through the evaluation reports of many of the developmental programs, one thing that is mentioned in all the reports in common is, the extension professionals were loaded with the technology, but lack in skills/technology of extension. Means the extension technology is missing, but the "technology" *per se* which is associated with the crop production or the animal production or the issues associated with, are dominating. Lack of extension technology needs to be taken care of, wherein the extension educationist is going to play key role. The role of extension service provider is, in addition to the technology that they are specializing with learn techniques of extension and influence behavior of the end user. Technology per sec cannot work on its own and do any wonders until and unless it is understood by the end user.

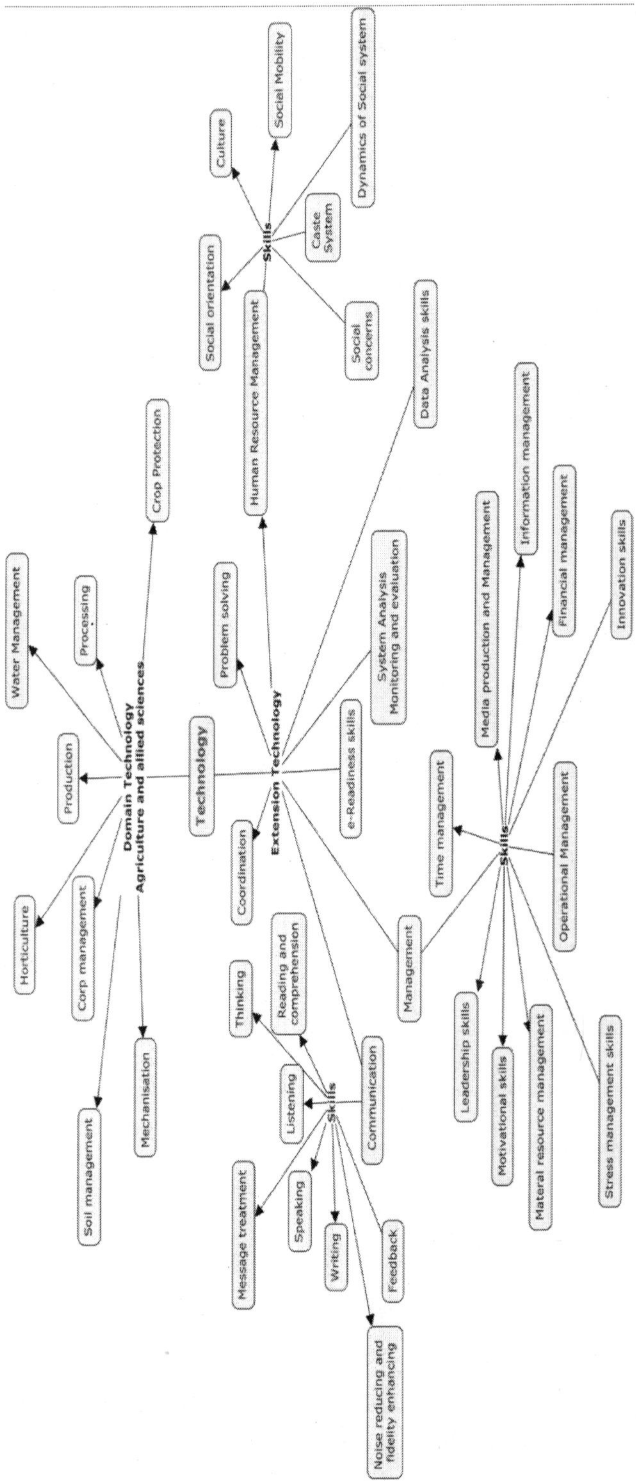
Technology
Domain Technology
Agriculture and allied sciences
Water Management
Processing
Crop Protection
Production
Horticulture
Corp management
Soil management
Mechanisation
Extension Technology
Problem solving
Human Resource Management
Skills
Social orientation
Culture
Social Mobility
Caste System
Dynamics of Social system
Social concerns
Data Analysis skills
System Analysis Monitoring and evaluation
e-Readiness skills
Coordination
Management
Skills
Time management
Media production and Management
Information management
Financial management
Innovation skills
Operational Management
Leadership skills
Motivational skills
Materal resource management
Stress management skills
Communication
Skills
Thinking
Reading and comprehension
Listening
Message treatment
Speaking
Writing
Feedback
Noise reducing and fidelity enhancing

Extension Technology

The emphasis of extension educator is on empowering the stakeholders, which can be understood with the saying, *"If you give a fish to a hungry person, you are satisfying his hunger for a day, but if he is being taught how to catch the fish, it means we are feeding him for the life"*.

The process of extension does not stop there. The inner-tensions of extension go beyond that. What is that? It says that, teach them how to process and package the fish for export, which stimulates the economic development. Bringing in the desirable changes is being emphasized upon with the inner-tensions of extension.

Extension educationist and extension service provider are different entities and have a definite role to play. Apply the techniques of extension while providing the extension services in addition to the technology which is essential for our target community, and make the process of extension more holistic. 'What is extension' is discussed everywhere but what extension is not about is not discussed anywhere. There are many misconceptions regarding the concept of extension. Because of these misconceptions, people don't understand holistically extension as a profession and as a discipline.

What is Not Extension?

The perception of people who don't understand the concept of extension say that extension is nothing but a type of gossip, is not a science, not for intellectuals, not academic discipline *etc.* Since beginning it is being emphasized on that it is the service as well as a profession. Because it is a service, it has developed as profession. Since it is academic discipline, it has its own subject matter, contents, syllabus and offers postgraduate degree programs. Some say that it is nothing but the act of photography, yes; extension makes use of the appropriate photographs, videos and everything for the benefit of the stakeholders for ensuring efficient services.

Extension education is not the process of story writing and it is also said extension education is a type of journalistic process. Developing success story and making use of few aspects of journalism in extension processes and models does not make them the profession of journalist. These misconceptions emerge because of the meaning attached to the concept of extension as the transfer of technology. There is another connotation attached with extension is the postman's job of transferring the information from resource institution to that place of information deficit. In between there are good number of educational events that are part of the extension service efforts, it proves to be not the postman's job because, until and unless there is change in knowledge, attitude and skill, we cannot say that extension has taken place.

Also few people say that it's a vague, hazy, indistinct and a fuzzy subject. People perceive it because they feel it is difficult to understand the processes of extension. It is difficult to understand all the nuances of extension. Hence it is not easy to prove, to be a professional extension educator or service provider. Extension deals with the most dynamic animal on this earth; that is human being, who is having competencies in various specialized areas. Extension Professional makes use of all techniques to convince every stakeholder. If he fails, then he starts feeling that, it's a vague subject, fuzzy subject, an indistinct subject.

Next misconception in the line is extension means crowd pulling or bringing the people to programmes. Obviously we need the people to understand the technology, to understand the information which is appropriate for them. It is not that sole responsibility of extension professional is bringing the crowd of people for some other purposes. The purpose is educational activities, for the purpose of changing their behavior, making them educated, changing their practices *etc.*

There is another understanding of extension; it is nothing but a propaganda activity. What do we mean by propaganda? Propaganda is the information especially biased or misleading in nature, used to promote a political cause or a point of view. When this concept 'propaganda' itself is having so many negativities within it how can we equate it with extension which has nothing to do with these negativities.

One more meaning attached with is extension means advertising, basically, advertising is an audio or visual form of marketing communication that employs an openly sponsored non-personal message to promote or sell a product, service or an idea. Extension deals with the information and innovative ideas, but they are not non-personal in nature. Extension does not market or sells ideas or innovations. But efforts are directed towards changing behavior, practices, living standards and ultimately leading towards enhanced production and productivity of its stakeholders.

Extension is not the process of mass communication. Because mass communication is a study of how people exchange information through mass media to large segments of the population at the same time. The process of extension deals with a particular target community. The information needs of target communities of extensions service are totally distinct. But in case of mass communication they are dealing the entire community as a whole. On the contrary extension segments the entire population based on their needs and provide information which is relevant to them. If you take a village as a unit, within the village, there are different information needs that the farmers have. There might be 10-15 groups of people who are in need of different types of information. Accordingly extension provides services to them.

Extension Educator Vs Extension Service Provider

What are the difference between the person who is acting as an extension educationist and who acts as an extension service provider? These are the people who are responsible for providing the extension services for the benefit of the people, hence there is need to overcome the misconceptions that we are trying to discuss about.

Extension educationist is a person with a background of education in agriculture and allied sciences and specialization in extension science and provides the knowledge and skills of extension and communication tools and techniques to the extension service providers at regular intervals to bring the desirable changes in the behavior of extension service provider. This is the role of an extension educationist who has specialized himself in extension science.

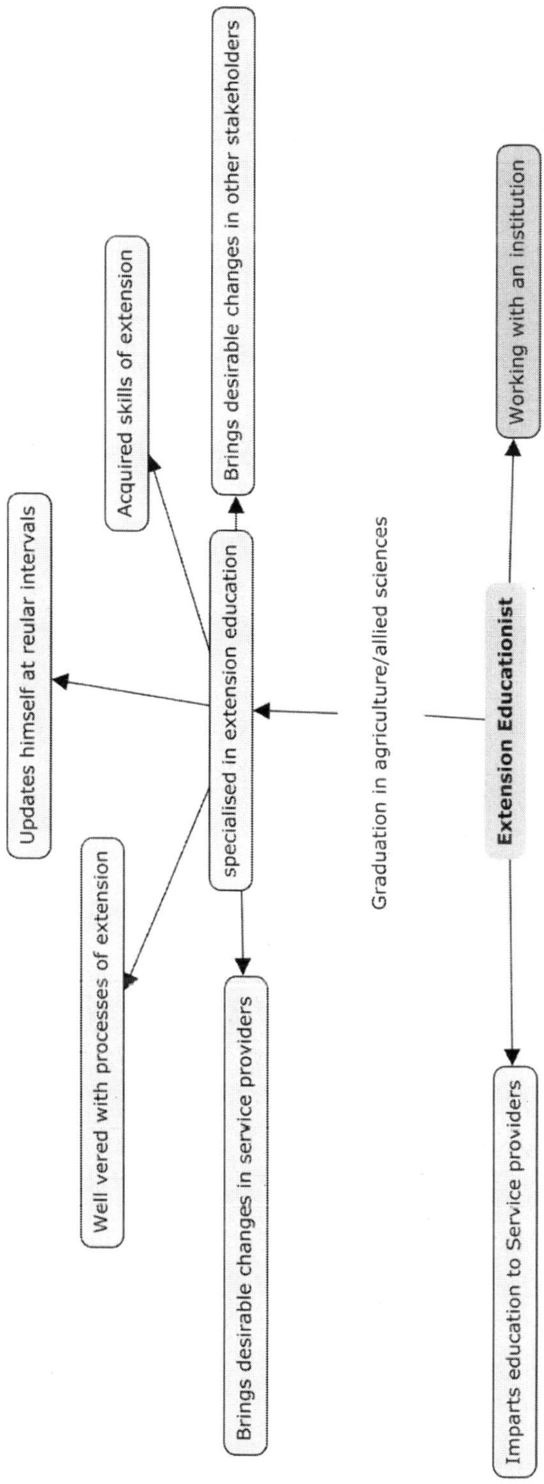
Extension Educationist
Working with an institution
Imparts education to Service providers
Graduation in agriculture/allied sciences
specialised in extension education
Brings desirable changes in other stakeholders
Acquired skills of extension
Updates himself at reular intervals
Well vered with processes of extension
Brings desirable changes in service providers

Who are the extension service providers? The service provider is a person with a background of education in agriculture and/or allied sciences and specializations in the discipline of agriculture and allied sciences, who is serving in a public or private institution and meant for dissemination of technological advances to the intended communities and institutions based on the needs, after acquiring the extension and communication tools and techniques from extension educationist.

The extension service provider is an individual with the specialization of his own subject, in addition to that, he is getting the additional information of how to transfer that technology from extension educationist, and then only he becomes service provider. For example entomologist, pathologist, agricultural engineer, agronomist, horticulturist, animal husbandry *etc.*

Extension Educators and Extension Service providers are responsible for transfer of information based on their specialization for the benefit of the farmers. Understanding these two terms, that is extension educationist and service provider, we can overcome the misleading concepts regarding the extension.

Pluralism in Agricultural Extension

The term pluralism means the existence of variety of agencies, service providers, models and institutional arrangements; specially including the public institutions, the private institutions, the community based organizations, the non-governmental organizations, catering to the information and advisory support and service needs of the farmers.

The institutional mechanisms are available for satisfying the information needs. When the number of such institutions are increasing, the concept of pluralism comes into the picture. It is not one institution but many institutions are providing the similar services. Why such situations are emerging especially in Indian context? Because the population dependent on agriculture is huge. Even today more than 50 per cent of the population is still dependent on agriculture. And the number of service providers to the population needs to be huge, but various players are interested in providing similar services.

According to the statistics available in the State of Andhra Pradesh, one extension worker is catering to the needs of 3,162 farmers, and when you look into input supply agencies, specially the fertilizer, which is one of the critical input for agriculture, for 1622 farmers, there is one fertilizer shop. Rest of the follow the suite. What does this mean? It means there is need for multiple institutions (pluralism) catering to the needs of extension services.

Strengths of Pluralism

- The diversity in approaches and the target groups covered, so that we can easily reach our clients in a comfortable manner and their needs are also satisfied.
- Can complement and supplement the efforts of each other. If the public extension mechanism is lacking in any of the efforts, the private can take care of that and if private is not able to do that, public extension agencies

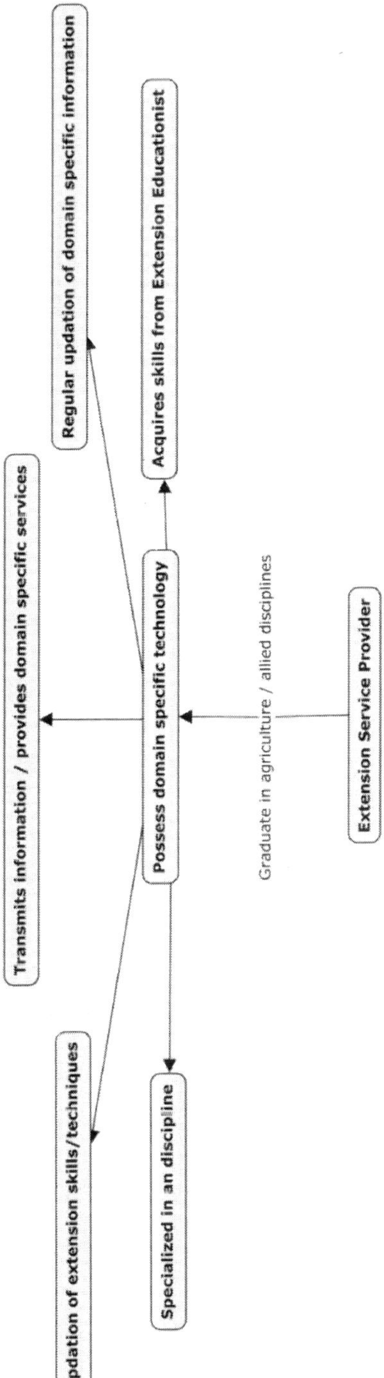
Extension Service Provider
Graduate in agriculture / allied disciplines
Possess domain specific technology
Transmits information / provides domain specific services
Regular updation of domain specific information
Acquires skills from Extension Educationist
Regular updation of extension skills/techniques
Specialized in an discipline

can do that, if both are not able to take up some issues the community based organizations or the non-governmental organizations can take into account.

- ☆ The private agencies can cater to the demands of the farmers, prominently observed in case of seed, plant protection chemicals *etc.*
- ☆ Help to overcome the paucity of human resource in the public extension system which is one of the major issue. Once the private players, the non-governmental organizations or the community based organizations are there, the paucity of the extension professionals can be easily taken care of. This is what is the strength of pluralism.
- ☆ It can help address varying needs of the farming community. The public extension mechanism (any such institution) may not be having expertise in all the areas as desired by the farming community. But the private institutions are coming out with their own super-specializations and accordingly they are catering to the needs of the farmers.
- ☆ Increase in outreach where public extension system is weak. There are many places where the public extension system is very difficult to reach, specially in the hilly areas, in the desert areas, in the States of Jharkhand or Uttarakhand or the North-Eastern States, where the public extension mechanisms are relatively weak, wherein the private as well as the community based organizations are taking care of such issues.
- ☆ Provision of newer technologies and skills to the farmers. So the private agencies can easily cater to these issues.

Weaknesses

- ☆ There is a possibility of duplication of effort, which the public institutions are also taking up and at the same time the private institutions are also taking up the similar efforts.
- ☆ Possibility of contradictory information flow and the resultant gap among the end-users. Specially we need to be very careful about the information flow from the private agencies, because their motto is the maximization of the profits. Then there are possibilities of creating some hype regarding the products. Such things we cannot expect from the public agencies. The validity and reliability of the information is being checked by various agencies, which is largely absent in case of the private agencies.
- ☆ Problem with the quality of the message from the public agencies. Whether we are satisfying the demands of the farmers or our clients is one of the issue, which is largely very weak in case of public extension system.
- ☆ Motives and objectives of all agencies may not be aligned with the needs of the farmers. Farmer is in need of something else, but as a private agency or as a community based organization or as a non-governmental organization my priorities are something different. So under such circumstances, there are certain gaps that might be existing between these service providers.

- It makes the extension system more complex, because of number of players that are coming on the same platform and making use of the same concepts of extension. So that is how the concept of pluralistic extension is becoming bit complex

Opportunities

- Harness the synergy of all the actors through convergence. The problems and the weaknesses can be addressed once we come on a common platform. This is an opportunity that we can think of, and address the issues and we can sort out the issues.
- Delineation of the roles of each agency or actor to avoid the duplication of efforts. If various agencies are on a common platform, we can resolve these issues and avoid duplication.
- Offering choice to the farmers among different agencies. So that lets farmer choose whether he wants the services from the public agencies or a private agency or a community based organization or an NGO so that we can provide them services accordingly.
- Enhancing the participation in case of extension programs is one of the opportunity, because as on date the real participation of the farming community in the extension programs is relatively less, but with pluralism we can enhance this participation.
- Public extension can play the role of a leader of pluralistic extension system. Because being part of the government mechanism, they can take the establish coordination among various agencies.
- Potential to create a linkage of the farmers with other actors and across the value chain. So that once the government takes the lead or the public institutions take the lead then automatically things can take a different shape. This is an opportunity.

Threats

- Lack of effective coordination among the agencies. Which is currently being experienced and over a period of time it may emerge as a threat also.
- Absence of platform for coordination. Some efforts are being carried out through ATMA. But still lot needs to be done in this sector.
- Efforts of collaboration without imbibing the spirit. Observed with the non-governmental organizations as well as the private sector organizations. The spirit of extension should not be compromised, while providing the services.
- Mismatch between the organizational agenda and the problems due to the hidden agenda. Generally we observe such things in case of the private agencies. Their agenda is profit maximization, while the agenda of the public institutions is the development of the society, enhancing the living

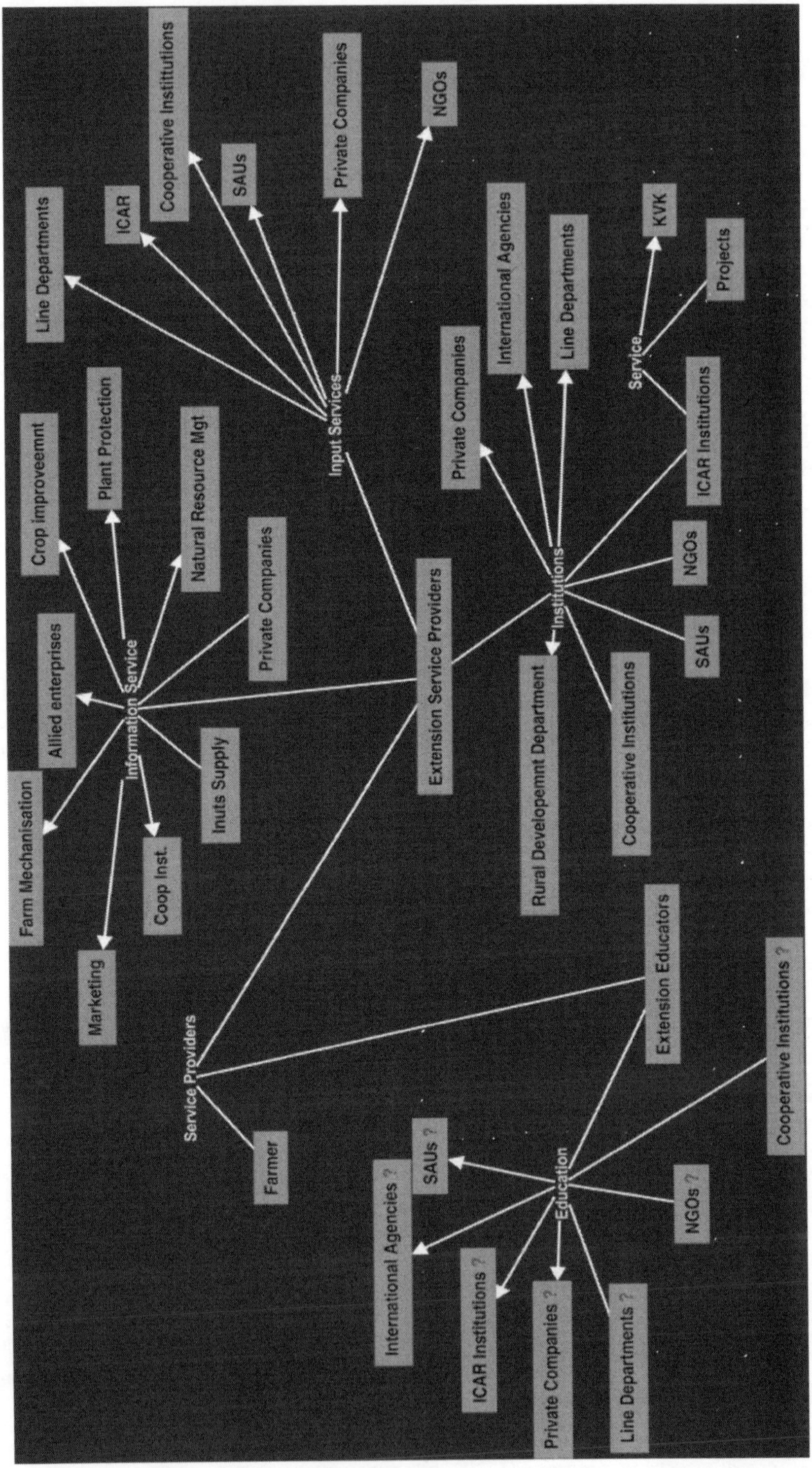
Extension Service Providers
Information Service
Crop improveemnt
Plant Protection
Natural Resource Mgt
Private Companies
Allied enterprises
Inuts Supply
Farm Mechanisation
Coop Inst.
Marketing
Input Services
Line Departments
ICAR
Cooperative Instittutions
SAUs
Private Companies
NGOs
Institutions
Private Companies
International Agencies
Line Departments
Service
KVK
Projects
ICAR Institutions
NGOs
SAUs
Cooperative Institutions
Rural Developemnt Department
Service Providers
Farmer
Extension Educators
Education
SAUs ?
International Agencies ?
ICAR Institutions ?
Private Companies ?
Line Departments ?
NGOs ?
Cooperative Institutions ?

standards of the farmers. There are obvious possibilities of mismatch between these agendas, the hidden agendas as well as the declared agenda.

- ☆ Lack of leadership and the conflict resolution mechanism may hamper the effective collaboration. Because when these institutions are to come together, even though their target groups are same, their commodities are same, the way of delivering the mechanisms, all are common. But whether these institutions are ready to come together or if there are any conflict that exist, do we have that conflict resolution mechanism? This is the biggest issue which is acting as a threat.
- ☆ Political and economic aspects preventing the collaboration. Because the motto of the private agencies and public extension agencies are diagonally opposite, is one of the hindering factor.
- ☆ Doubt about the sustainability of the government promoted convergence initiatives is creating another threat in case of pluralism in extension.

If you look into this diagram, the services we are providing to the client *i.e.* farmer. He needs the services. The services can largely be identified as the services to be provided by the extension educators and extension service providers. So when we look into the extension educational efforts, they are being taken up by the State Agricultural Universities, the international agencies, the ICAR institutions, the private agencies, the line departments, the non-governmental organizations, cooperative societies number of institutions which are involved in educating the farming community.

Similarly when it comes to the service, the institutions which are involved in the service may it be the department of rural development, the cooperative institutions, the state agricultural universities, non-governmental organizations, ICAR institutions, KVK's as well as the projects, they are reaching the farmers. The line depts, international agencies and the private agencies are also providing services.

The institutions discussed under the extension education as well as extension service are almost same. This is what is leading towards the pluralism. The input services are being provided again by line depts., ICAR institutions, cooperatives, state agriculture universities, private companies, NGO's. They are all involved in the inputs and at the same time the information services, which is again the domain of the public extension agencies. The private players are also coming here, the similar institutions are there. So this is what we are emphasizing on. Similar services are provided by a number of institutions, ultimately the farmer is getting confused; which kind of services is to be accepted and which kind of services not to be accepted.

Role of ICTs in Pluralistic Extension Services

ICTs have great potentials to cater to the needs of pluralistic extension services. Before analyzing the potentials lets look into some interesting facts about websites.

- **1,518,207,412 (1.5 billion) in the World as of January 2019**
- **Of which only 15% are active**
- **More than 85 per cent sites are mobile friendly**
- Google has indexed 4.5 billion web pages
- Every minutes 380 and everyday 5,47,200 websites are created
- Of the 700 living languages websites are found in 200 languages
- Language and website share
  - English – 51 %
  - Russian – 6.7%
  - German – 5.7%
  - Japneese and Spanish - 5.1
  - French – 4.1
  - Others – 21.1 %

https://www.millforbusiness.com/how-many-websites-are-there/

There are about 1.5 billion websites available by January, 2019. So this shows the number of service providers specially in case of with e-mediation, apart from the traditional medium what we have discussed so far. If you look into the functioning of these websites, only 15 per cent of them are active, 85 per cent of them are not active over a period of time. Out of this more than 85 per cent of the websites are mobile friendly. It means this is an opportunity for service providers, as well as the service receivers. Google itself has indexed about 4.5 billion pages on an average 3 web pages per website that we can say so it has indexed. The knowledge repository is getting ready. But the service receiver also should mend himself to get these services.

While, if you look into the information that is available in the language wise. English is dominating. But in case of India so there are multiple languages that are being spoken. None of the Indian languages are dominating the websites. This has implications on Indian extension mechanism that, there is a need for development of the content in the local as well as the regional languages, otherwise we cannot provide the appropriate information to the end user.

e-mediation promotes pluralism, the number of service providers make use of this e-mediated platform by developing content in the regional as well as the location specific information. The associated risk factors need to be taken care of by the institution itself and at the same time the receivers are also playing a role by accepting that information. Then extension professionals face challenge in content development suitable for the e-mediated extension, because there is a lot of information that is available, but now it needs to be treated for suiting to the needs of the farming community. So that is where the role of extension is coming. Again we can make use of the number of players and ultimately we can provide them the appropriate services.

Indicative List of Efforts of e-Mediation in Extension Services by Various Institutions

e-Extension Initiatives of ICAR

- Rice Knowledge Management Portal (RKMP)
- Consortium for e-Resources in Agriculture (CeRA)
- KRISHI
- Mobile apps (Herbal kisan, Mushroom app, Farm Calculator *etc.*)
- SMS services by KVKs
- National Initiative on Climate Resilient Agriculture (NICRA)
- Design of Micro Irrigation Systems (DOMIS)
- e-Ddalhan gyan manch
- Knowledge Innovation Repository of Agriculture in the North East
- Expert System for Maize
- Expert System on Wheat Crop Management
- Expert System on Seed Spices
- CaneInfo
- Technologies and Products for Commercialization -Animal Science

e-Extension Initiatives of SAUs

- TNAU Agritech portal
- KAU Agritech portal
- Pau Kisan app, Seed portal
- GADVASU pigfarming app, goat farming app
- Bihar krishi app, Bihar agrodoctor

e-Extension Initiatives of NGOs

- DHAN foundation ICT enabled CRS
- MSSRF – Fisher friend app
- Nandi foundation – Facebook page
- AFC India Ltd – Customised farmers online training
- Digital Green – Video content creation in regional dialests
- Jayalakshmi Agtotech – Crop specific mobile apps in regional languages

e-Extension Initiatives of Private Sector

- ITC e-Chaupal
- TATA kisan Kendra
- Hariyali Kisan Bazar

- ✰ Krishiworld.com
- ✰ Myagriguru app
- ✰ Agriwatch.com

The efforts need to be intensified to reach the unreached, the reached needs to be satisfied and the satisfied need to be brought to the net of extension delivery mechanism. What should be role of extension educationist in e-Mediated extension service delivery. There is no need to arrange for crowd, public address system *etc.* again overcoming the misconceptions and understanding the role of extension is key issue here.

An Extension Educationist should

1. Understand the pulse of stakeholders
2. Prioritize the needs of stakeholders
3. Educate the extension service providers
4. Make extension educators to organize the contents as per the needs of stakeholders
5. Treat the content developed by extension service providers
6. Fine tuning of the content
7. Putting content into various e-Mediated platforms
8. Develop mechanism for regular and continuous feedback mechanism
9. Evaluate the feedback on regular and continuous basis
10. Reconsider the content and add value to the content and present it for the stakeholders

As soon the first cycle of operations' are over lets begin with the first step again and the process of extension education continues.

Apart from the above listed efforts of providing the services to stakeholders and educating service providers, the innovative method of educating the stakeholders of extension is through offering Massive Open Online Courses. One such effort of providing education to various stakeholders is being made through agMOOCs platform. As we are aware that agricultural education in India is offered through a network of 75 State Agricultural Universities, five Deemed Universities, four Central Universities with faculty of Agriculture, three Central Agricultural Universities and few privately owned colleges of agriculture affiliated to traditional state universities. Each Agricultural University has a number of constituent colleges. However, paucity of quality faculty has remained an emerging issue. Specialization of the faculty also matters while imparting education. Providing quality inputs to the learners spread over large geographical area demands e-Mediation.

One such platform was created by Commonwealth of Learning (COL) and Indian Institute of Technology, Kanpur entitled "agMOOCs" in 2015.The courses offered through the platform have attracted participants from every state of India and seven countries. The paper is a partial analysis of the efforts of agMOOCs in penetrating agricultural education system as an alternative platform for students of agriculture and allied sciences. More than two lakh learners have accessed courses so far. The feedback of learners is highly encouraging. So far the authors have offered three courses in agMOOCs platform during 2017 and 2018.

The highest number of learners registered in a course was 4884. There is an increasing trend in number of registrations on the platform, which shows effectiveness of the courses and penetration among agricultural fraternity. Age of majority registered learners was up to 24 years (76 per cent). More than 65 per cent offered courses to enhance their knowledge. Home was the most preferred place of access (65 per cent). About 60 percent learners preferred the agMOOCs app to access course contents. Gender and preference of app are not independent of each other, Gender and age of participants are not independent, there was no association between range of learners and relevancy of quiz. There exists association between range of learners and pace of content delivery (adopted from: http://oasis.col.org/bitstream/handle/11599/3263/PCF9_Papers_paper_40.pdf?sequence=1 and isAllowed=y.

Conclusion

The results of the effort indicate that there is growing inclination towards user friendly platforms. Given an opportunity stakeholders are participating and getting the benefits, but how many such programs are being offered? How to ensure the quality aspects of content? How to treat the message as per the needs to stakeholders? How to document and interpret the needs of stakeholders? Who and how to decide the appropriate platforms for various stakeholders? And so on. The list of questions to be answered is growing, its only the extension educationist. This is the opportune time to extension educationist to recognize his role and begin with the educational activities of extension service providers and other stakeholders.

REFERENCES

Aditya and Jirli, B. 2010. ICT mediated agricultural extension: a survey of leading models and best practices, *Journal of Progressive Agriculture*, 1(1): 74-83.

Aditya and Jirli, B. 2011. A study on social computing aspects on the students of Banaras Hindu University *Journal of Global Communication*, 4(2): 148-161.

Aditya, Thakur, A.K. and Jirli, B. 2014. Awareness about Social Computing among students. *Asian Journal of Extension Education*, 32(1): 21-23.

Aditya, Jirli, B. and Thakur, A.K. 2014. Values Associated with the Students Using Social Media Tools. *Journal of Community Mobilization and Sustainable Development*, 9(1): 93-96.

Aditya, Kumar, B., Abhinav, K. and Jirli, B. 2014. Technology enabled learning in South Asia: A Review. *Journal of Global Communication*. 7(2): 128-134.

Ghatawal, J.P. and Jirli, B. 2016. New Media activism among young learners, *Journal of Global Communication*. 9(special issue): 355-366.

Ghatawal, J.P. and Jirli, B. 2016. Communication And Psychological Profile Of Central Office Employees of Banaras Hindu University. *International Journal of Agriculture Sciences,* 10(12): 6459-6461

Ghatawal, J.P., Jirli, B. and Singh A.K. 2016. Recognition and Utility of New Media for Students. *International Journal of Agriculture Sciences,* 8(53): 2761-2764.

Jirli, B., Srivastava, S.K. and Singh, R.S.P. 2006. A Study on attitude of undergraduate students towards distance education in agriculture. *Journal of Open Schooling,* 4(1): 56-71.

Mahara, G.S., Jirli, B. and Rai, A. 2013. m-Learning: An educational approach for Change in Hills of Uttarakhand. *Journal of Global Communication.* 6(2): 144-150.

Mandal, P.K., Kirti and Jirli, B. 2016. Role of Information and Communication Technologies in Rural Development *Journal of Global Communication.* 9(special issue): 129-133.

Meena, R.L., B. Jirli, M. Kanwat and Meena, N.K. 2018. Mobile Applications for Agriculture and Allied Sector. *Int. J. Curr. Microbiol. App. Sci.* 7(02): 2317-2326.

Rai, A. and Jirli, B. 2013. Information Technologies for e-Readiness. *Journal of Communication Studies.* 31: 28-33.

Rai, A., Jirli, B. and Singh, A. 2016. e-Readiness 4D: An Analysis. *Journal of Global Communication.* 9 (1): 22-31.

Rai, A., Jirli, B. and Singh, A. 2018. e-Readiness: Its Relational Perspectives with Profile of Extension Functionaries. *Journal of Global Communication,* 11(1): 65-75.

Singh, A. and Jirli, B. 2018. Changing Perception and Awareness towards ICT along with Changing Roles of Stakeholders in Agriculture. *Research Journal of Agricultural Sciences,* 9(6): 1412-1416.

Chapter 10

Development and Field Testing of Need Based Mobile App on Buffalo Reproduction: An Educational Tool for Information Dissemination

Hema Tripathi[1], *Vishwa Bharati Dixit*[2], *Sajjan Singh*[3] *and Dharmendra Kumar*[4]

[1]Principal Scientist, Transfer of Technology Unit,
[2]Principal Scientists and Incharge Transfer of Technology Unit,
[3]Principal Scientist, Animal Reproduction Division,
[4]Senior Scientist, Animal Reproduction Division
ICAR-Central Institute for Research on Buffaloes, Hisar – 125 001, Haryana

Identification of needs is one of the prerequisites for development of ICT tools and strong information delivery system for smoother and faster access of information for farmer's doorstep for receiving and sharing scientific information and knowledge. Properly addressed and most needed information can act as an input to sustain a production system for quick and accurate decisions. Like other production systems, farmers have specific needs in buffalo production system too, to adopt good buffalo farming practices.

Amongst the broad areas, buffalo reproduction plays a great role for sustainable livelihood and welfare of buffaloes. Reproductive and production disorders of buffaloes significantly reduce their productivity which is of great concern of dairy producers worldwide because most reproductive disorders adversely affect the future fertility. The major problems commonly observed in the field have direct impact on reproductive performance of buffalo are; abortion, dystocia, retained placenta, metritis, prolapse, anoestrus and repeat breeding. Ten to 30 per cent of lactation may be affected by infertility and reproductive disorder and 3-6 per cent of the herd is culled annually in developed countries for these reasons. Studies have shown incidence rate of uterine infection, incidence of endometritis, incidence of anoestrous, repeat breeding incidence of abortion and still birth dystocia, and retention of fetal membranes are higher in buffaloes than cows. Poor reproductive performance reduces the number of calves born and milk production and may increase the cost of therapy and semen. Prolonged postpartum anoestrus, lack of knowledge on oestrus detection and early pregnancy diagnosis are the factors that limit the efficiency of buffaloes and ultimate economic loss to the owner caused by decreased production and delayed reproduction as well as increased treatment and preventive measurement costs due to lack of awareness and knowledge on improved scientific practices.

Most of the livestock owners in India lack required information on buffalo reproduction practices due to poor information delivery mechanism especially in animal husbandry, which otherwise could enhance buffalo productivity and product quality.

The contribution and population of buffaloes in India (109.85 Million) is showing an increase of about 1.0 per cent over previous census. Female buffalo population has also increased by 8.61 per cent whereas male buffalo declined by 42.35 per cent over previous census. In our country, wherein about 20.5 per cent of the total livestock is contributed by buffaloes (20th livestock census) is our duty to strengthen our existing information delivery system to reach the buffalo owners to the deepest and remotest place with the aim to acquaint them with latest scientific information without delay to support the production. Lack of information has been cited as one of the most important factors for dairy product quality vis-a-vis prescribed standards in India. Formal extension institutions currently seem unable to meet the information and skill requirements of vast segment of dairy farmers. Dispersed rural populations; high cost of delivering information in person and a lack of accountability have limited the efficacy of traditional extension systems in India). Under the circumstances wherein majority of buffalo farmers approach non-formal sources for information on scientific buffalo farming, sometimes mis-information leads huge loses to farmers as well as to animals also. Thus providing right information at right time from credible source is the need of today to enhance the individual farmers herd productivity.

Harnessing the Power of ICT in Livestock Sector

ICT can be the best solution as it plays an important role in timely information access and delivery. Among ICTs, a mobile based application is gaining a huge popularity and momentum now-a-days in every sector because of their ease in access and faster dissemination of information in both in rural and urban areas and has greater rural penetration than any other ICT. More and more smallholder farmers are now harnessing the power of the mobile telephones in connecting themselves to various services, information and markets. International Dairy Federation recommends that dairy producers need to have information about good dairy farming practices to support the production and marketing of safe, quality-assured milk and dairy products. Thus use of need based mobile apps will certainly help in gaining knowledge, problem solving and adoption of scientific practices by the buffalo owners for augmenting the production and productivity. With these benefits, Mobile apps are much focused in dissemination of information pertaining to buffaloes. The present research study was undertaken to answer the basic research questions like; what are the information needs of the farmers with special reference to buffalo reproduction system and to what extent the developed mobile app would be perceived as useful to the different stakeholders for adoption and further application?

Keeping the above in view, a study on development and field testing of need based mobile app on buffalo reproduction was undertaken that could be used as an educational tool for information dissemination. Before development of content of app, information needs of buffalo owners were assessed along with their accessing, sharing and utilization pattern of information at first instance on various facets of buffalo reproduction. After development of app, data were also collected from different stakeholders regarding the perceived utility of app for adoption.

For the purpose of research, Haryana and Rajasthan states were chosen purposively due to accessibility and ease in communication. Data has been collected after developing bilingual interview schedule to understand the information needs of buffalo owners in 10 adopted villages under *Mera Gaon Mera Gaurav* (MGMG) program from two different states from farmers having more than 02 adult milch buffaloes on three point continuum ie Most needed, Needed and Not needed with respective scores of 3, 2 and 1 in the area of buffalo reproduction and related parameters besides on socio economic characteristics of the buffalo owners. Data so collected was tabulated and scored against each of the predefined activity in each areas and the ranking was done based on mean weighted scores.

Information Needs of Buffalo Owners in Buffalo Reproduction System

For identification of needs of buffalo owners to develop need based mobile app, data were collected on ten parameters like; age at puberty and maturity, heat symptoms and heat detection, artificial insemination, pregnancy diagnosis, gestation period, inter calving period, age at first calving, castration of scrub bulls, peri partum care, bull selection, management and use. Out of these 10 parameters, heat symptoms and heat detection got the highest rank followed by age at puberty and maturity and peripartum care in descending order. Information on gestation period, pregnancy diagnosis castration of scrub bull was the least required needs revealed by the respondents.

Under buffalo reproductive problems, data were collected on 16 different problems of buffalo reproduction. These were; anoestrous, dystocia, infertility, abortion, retention of placenta, delayed puberty, repeat breeding, torsion of uterus, prolapse of uterus, metritis (infection), pyometra, sterility at early age, delayed age at first calving, venereal diseases, fetal maceration and mummification. Out of 16 problems; infertility, prolapse of uterus and retention of placenta were the most

needed information areas and got the highest rank followed by anoestrus, abortion and dystocia in descending order. Information on vaginitis, sterility at early age, pyometra and delayed age at first calving were the least required information needs revealed by the respondents under reproductive problems.

Development of Need Based Mobile App on Buffalo Reproduction

Buffalo Reproduction app is an educational mobile app designed and developed based on the needs of the farmers under buffalo reproduction in Hindi and English languages with aim of imparting knowledge to buffalo owners and aguide for VLDA and graduating veterinarians. The App provides basic information on different areas of buffalo reproduction by dividing the whole content into seven distinct segments. The major areas covered are; targets of reproduction, puberty and sexual maturity, heat symptoms and breeding methods, pregnancy diagnosis, peripartum care and management, bull selection and management for breeding. Special emphasis has been given by covering eleven major reproductive problems commonly found in buffaloes emphasizing on causes, preventive and control measures on each problem/disorder separately. It includes delayed puberty, anoestrus and silent estrus, repeat breeding, genital prolapse, dystocia, retention of placenta, cystic ovaries, venereal diseases, abortion, fetal mummification and fetal maceration with unique photographs. The App additionally provides answers on frequently asked questions by the farmers under each section of buffalo reproduction and complete App content has audio backup with download facility. After placing the app on Google Play store.

The link of the Buffalo Reproduction App is https://play.google.com/store/apps/details?id=com.cirb for smooth download.

Field Testing of Developed Mobile App on Buffalo Reproduction

The developed appwas field tested for assessing the effectiveness and perceived utility, using an interview schedule from 50 farmers, 50 stockman/VLDAand 50graduating veterinary students after their downloading of app, for further modification based on their judgement. The data revealed that about 83 per cent respondents had high readiness, motivation, innovativeness and trustworthiness for accepting the app. Only 36.7 per cent had prior experience of using the app on mobiles and about 60 per cent farmers showed technophobia in using the app. More than 83 per cent respondents found the app as user friendly, attractive and reliable with 70 per cent agreed for ease in downloading. More than 65 per cent respondents found that content and photos used in app as interesting, informative with good designing. According to 60-70 per cent respondents,app had high video streaming with eye catching colours with easy visuals and audio streaming with moderate pauses and low voice modulation. Fifty eight-70 per cent respondents found the content of the app as appropriate, clear, updated whereas preciseness, relevancy and coverage were found moderate by about 45-50 per cent respondents.

Overall utility of the app perceived by respondents was found high as revealed by 70 per cent farmers, 80 per cent graduating veterinary students and 55 per cent VLDA's. Seventy five per cent of the farmers, VLDA and graduating students perceived excellent look of the app and were not found statistically different in their perception regarding the effectiveness and perceived utility of the mobile app with special reference to buffalo reproduction. Thus development of buffalo based mobile apps directly helping different stakeholders in the accessing timely, optimum and right information and could help in reducing the economic loss and wastage of efforts, resources, money *etc.* that directly influence on production and productivity of buffaloes.

More than 1000 downloads have been done by different stakeholder in the 06 months after uploading on Google Play Store that clearly shows the utility and significance of this app.

Significance of Developed Mobile App on Buffalo Reproduction

- ☆ Acting as an alternative extension service delivery and focussed advisory system for solving specific problems on buffalo reproduction by reaching greater number of buffalo owners living in remote areas by minimizing the transaction cost.
- ☆ In providing demand driven information rather than supply driven information.

- For use by vets, para-vets working in SDAH, NGOs, SVU, veterinary colleges *etc.* to perform various services effectively.
- As a training tool for the youth aspiring to become para-vets.
- In helping the graduating veterinarians to improve their skills on various facets to prevent them from committing mistakes in the procedure.
- As a very good tool for strengthening the doorstep delivery services by helping the vets and para-vets to refresh their knowledge.
- In imparting the scientific knowledge to buffalo farmers in local language and at the farmer's level of understanding.
- Easy in downloading on android mobile phones and can be consulted at any place and any time for improving the production and productivity of buffaloes.
- In saving money, time and effort and act as a source of readily available appropriate and need based information.

The findings and suggestions produced would be helpful to the policy makers and administrators to formulate strategies to strengthen veterinary extension system and information delivery system by the use of ICT tools specifically through mobile.

Outcome of Study

- Mobile app on buffalo reproduction is available at free of cost on Google Play store to the buffalo owners, graduating students, paravets and others engaged in buffalo development.
- It is a portable resource for buffalo owners and other veterinary professionals.
- Developed mobile app eliminated the time and distance barrier that get in the way of knowing the latest information on buffalo farming from any part of the world.
- This app is available and accessible on Google Play Store all the time, 24 hours a day and 365 days a year
- The perceived needs of the buffalo owners were taken care in developing the app to provide them demand driven and value added information through mobile app.

Chapter 11

Transforming Indian Agriculture through Digital Platforms: Status, Support Systems and Strategies

Lakhan Singh[1] *and Sunil Kumar*[2]

[1]*Director, ICAR-ATARI, Pune, Maharashtra*
[2]*SMS, KVK-Kolhapur-II, Maharashtra*

Prospects of Transforming Indian Agriculture

Agriculture and allied sector continues to be pivotal to the sustainable growth and development of the Indian economy. Not only does it meet the food and nutritional requirements of 1.3 billion Indians, it contributes significantly to production, employment and demand generation through various backward and forward linkages (*GoI, 2016*). Although global food production has increased in the past few decades, almost 800 million people still have insufficient food in South and Southeast Asia, the Caribbean, and Sub Saharan Africa (FAO, 2015). Moreover, global food production must double by 2050 to match population and income growth (Alexandratos *et al.*, 2012), and much of this must happen in Asia and Africa. In India, small and marginal farmers' average size of holdings is about 0.38 ha when compared to 17.37 ha for large farmers, which cannot generate adequate employment and income from crop cultivation (Dev, 2017). In India small and marginal holding farmers cultivate around 44 per cent of the area, and they produce around 60 per cent of the total food grain production (49 per cent of rice, 40 per cent of wheat, 29 per cent of coarse cereals and 27 per cent of pulses) and over half of the country's fruits and vegetables production (Agricultural census, 2014).There is growing demand for rapid input, service and information delivery among the farmers although, fulfilment of these demands solely by public extension system is limited (Mukherjee *et al.*, 2012).These millions of small and marginal farmers and many of those are illiterate and have little or no access to resources to access modern technology in agriculture (Yadav *et al.*, 2015). Information and Communication Technology (ICT) has emerged as a tool for achieving meaningful societal transformation and it is an emerging tool for achieving meaningful societal transformation (Meera *et al.*, 2004).

The following are some of the striking feature of Indian agriculture in recent years.

Declining Share of Agriculture in GDP

The share of agriculture (which also includes livestock besides crops) and allied sector (forestry and logging and fisheries) in aggregate GDP has declined steadily over the past several decades from as high as over 50 per cent in the 1950s to around 14 per cent in recent years. The key challenges faced at present by Indian agriculture on the production side are shrinking land base, inadequate water resources and the vagaries of nature, shortage of labour and increasing cost. With the growth rate in the agricultural sector likely to fall significantly short of overall economic growth of 6 to 7 per cent in coming years, the share of agriculture GDP to the overall GDP will continue to decline. If the current trend continues, agriculture's share in aggregate GDP may decline to less than 10 per cent by 2019-20 (Figure 11.1).

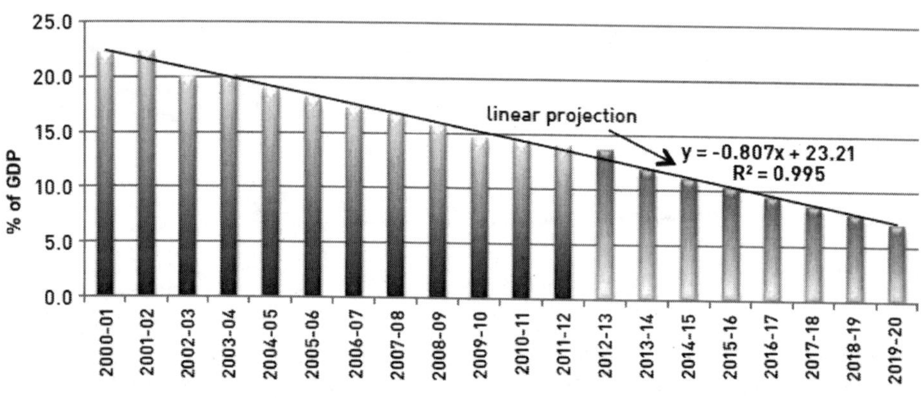

Figure 11.1: Agriculture and Allied Sector GDP Share in Overall GDP (*Source*: Fourth Semi-Annual Medium Term Agricultural Outlook Report, September, 2015, NCAER).

Shrinking Farm Size

A major challenge facing Indian agriculture is that the average-farm holding size is shrinking with the average farm size in 2010-11 estimated at 1.16 hectares1 compared to 1.23 hectares in 2005-06, and 2.28 hectares in 1970-71 (Agricultural Census 2010-11). The total number of farm holdings has almost doubled from 71 million in 1970-71 to 137.8 million in 2010-11. If this trend continues, farm holdings in 2020-21 would number around 154 million with small and marginal holdings accounting for almost 85 per cent of total holdings and the average holding size projected to decline to just one hectare (Figure 11.2).

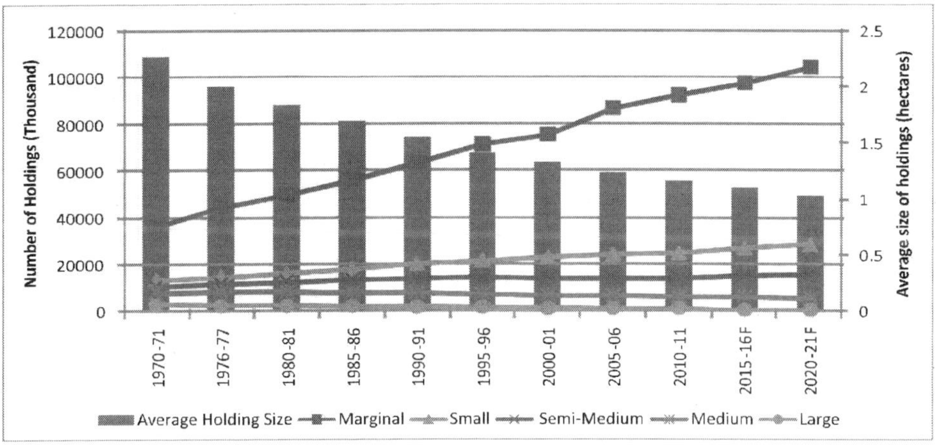

Figure 11.2: Indian Farm Holding Size Shrinking (*Source*: Fourth Semi-Annual Medium Term Agricultural Outlook Report, September, 2015, NCAER).

These facts indicate that the small holding farmers in Indian agriculture are much more prominent today than before. Major problems which have disturbed the income of small and marginal farmers during the last few years can be summarized as below (Hegde, 2019):

- Erratic climate change and lack of suitable technologies to cope with changing environment, thereby affecting crop yields.
- Poor irrigation facilities due to inefficient distribution of available water resources and slower progress of new irrigation projects. Interrupted supply of electricity further posed a hindrance in use of the available water resources for irrigation.
- Rising cost of inputs, small and fragmented land holdings and increasing labour costs have resulted in high cost of production and lower crop yields.
- In the absence of decentralized storage and processing facilities, small and marginal farmers could neither hold their produce till the realization of higher price nor process for value addition
- Poor agricultural extension and information services about selection of suitable crops, improved technologies, weather forecasts, types of pests and disease prevailing in the region, information on demand and price for various farm produce, *etc.* could not empower the farmers to take suitable corrective measures on time.
- Lack of required credit to procure various inputs well in time, high premium for crop insurance, and delay in settling the claims of farmers, created a cash crunch, beyond their ability to tide over the crisis.
- Low productivity of livestock high cost of feeding due to shortage of feed and fodder, inability to dispose of old and unproductive cattle and lack of clarity about the economics of dairy husbandry, created uncertainty about the future.

Type of Support Systems

1. Soil Health Card Portal

Table 11.1: State-wise Status of Soil Health Card Scheme (2015 to 2017) and (2017 to 2019)

Sl. No.	States/UTs	No. of Samples Tested (2015 to 2019)	No. of SHCs Distributed (2015 to 2019)	No. of Samples Tested (2017 to 2019)	No. of SHCs Distributed (2017 to 2019)
1.	Uttar Pradesh	4770399	16991346	4867442	18881000
2.	Maharashtra	2347121	12977232	2792042	13053011
3.	Madhya Pradesh	2313977	8872377	2315844	8907385
4.	Rajasthan	2308013	6886000	2865039	11106358
5.	Karnataka	1665765	7832189	1665766	7832189
6.	Gujarat	1589236	5108923	1675912	8571923
7.	Andhra Pradesh	1348382	7455204	1357706	6967162
8.	Bihar	1308778	6469650	1140373	5203849
9.	West Bengal	1300349	5040510	1167860	3089181
10.	Tamil Nadu	1274536	7000000	1300037	6974843
11.	Telangana	1034678	5720737	954453	3165530
12.	Punjab	835526	1251726	835526	1160568
13.	Haryana	788670	4227238	1245100	2932654
14.	Chhattisgarh	703691	3890709	949334	4746670
15.	Odisha	668635	2374233	579383	1291711
16.	Kerala	127585	763435	141454	2052653
17.	Goa	25000	25000	12000	12000
18.	Uttarakhand	135738	750494	137019	882058
19.	Himachal Pradesh	69635	385011	100000	960765
20.	J and K	160687	692062	147853	1018051
21.	Jharkhand	115302	637507	76005	252834
22.	Arunachal Pradesh	20532	20532	8005	0
23.	Assam	278707	1300901	284778	301864
24.	Manipur	20713	114522	13746	44516
25.	Meghalaya	39372	209561	39372	236378
26.	Mizoram	11986	11986	11986	14974
27.	Nagaland	33423	184797	16712	12000
28.	Sikkim	13217	46000	10109	15360
29.	Tripura	32736	117723	32736	117723
	Union Territories				
1.	Andaman and Nicobar	1405	10000	11804	8970
2.	Dadra and Nagar Haveli	2222	2222	0	0
3.	Puducherry	3530	19594	5747	7074

Source: Data retrived from https://soilhealth.dac.gov.in/

Soil is the key to ecosystem functions (Hartmann *et al.*, 2015), which support the production of food, forestry products and human health (Lal,1990). The SHC scheme has been approved for implementation during the remaining period of the 12th plan with the aims to provide SHC to all farmers in the country at a regular interval of two years to enable the farmers to apply recommended dosages of nutrients to realize improved and sustainable soil health and fertility status and to reap higher profits (Reddy, 2018).During 2015-2017 total 107389421 number of SHC has been distributed. During 2017 to 2019 till 1st October, a total of 109822114 have been distributed.

2. Improving Market Access

A. India is predominantly an agrarian economy, with agricultural sector engaging about half of the workforce (GoI 2016). About 48 per cent households in India are agricultural households, whose monthly income is Rs. 3140 from crop cultivation alone (NABARD 2018). Agricultural markets are real or virtual meeting places where buyers, sellers and intermediaries meet to exchange commodities between seller and buyer (Reddy, 2018). In the exchange process, sellers, buyers and intermediaries face many counter-party risks, like delivery failures, substandard quality and delay in payments. There is potential for opportunism given the divergent interests of different market participants (Williamson, 1985).

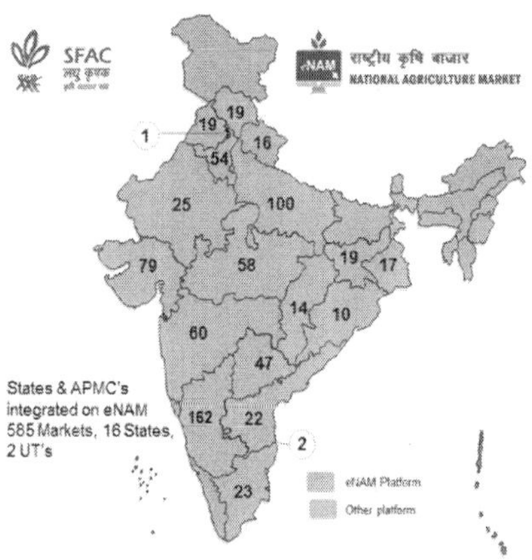

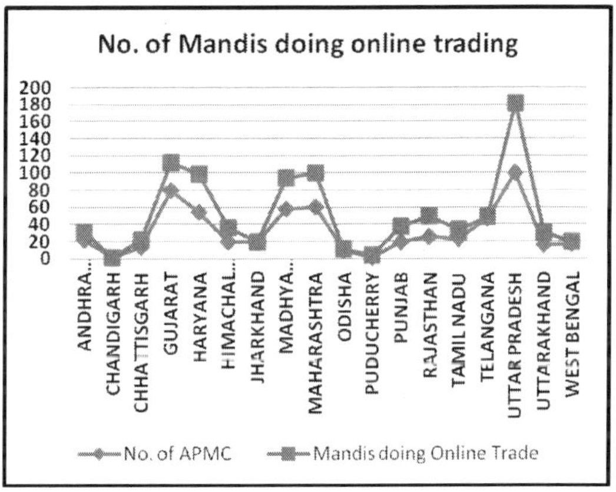

e-NAM have multi dimensions like to establish a national e-market platform for transparent sale transactions and price discovery initially in regulated markets, to provide liberal licensing of traders/buyers and commission agents by state authorities without any pre-condition of physical presence or possession of shop/premises in the market yard, to create one license for a trader valid across all markets in the state, harmonization of quality standards of agricultural produce and provision for assaying infrastructure in every market to enable informed bidding by buyers *etc.* (Bachaspati *et al.*, 2018).

Table 11.2: Status of Stakeholders in e-Nam

State	18
Traders	1,25,626
Commission Agents (CAs)	70,111
Service Provider	0
FPOs	826
Farmer	1,65,33,086
Total	**1,67,29,649**

Source: https://enam.gov.in/

B. **Agmarknet:** The AGMARKNET website (http://www.agmarknet.nic.in) is a G2C e-governance portal that caters to the needs of various stakeholders such as farmers, industry, policy makers and academic institutions by providing agricultural marketing related information from a single window. The portal has helped to reach farmers who do not have sufficient resources to get adequate market information. It facilitates web-based information flow, of the daily arrivals and prices of commodities in the agricultural produce markets spread across the country.

C. **Agri-tech startups:** Today, the agri-tech sector is witnessing a number of startups in India disrupting everything from organic farming and equipment rentals to connected supply chains and cloud-based analytics. Some of them are listed here:

   Farms2Fork offers farmers water monitoring solutions that ensure better productivity by reducing water wastage. The solution includes IoT wireless soil sensors, AI support, and real-time analytics. While earlier agri-tech solutions were based on batch processing of data, Farms2Fork operates on real-time data.

   Agribolo, founded in 2015, is a farming services platform spanning activities such as information dissemination, quality input procurement, market linkages, irrigation facilities and farming equipment.

   AgroWave, founded by an IIT Delhi alumna in 2017, aims to optimise agriculture supply chain using research, analytics, and technology. Demand and supply analytics connect farmers in Panipat, Sonipat, Harpur, and Rajasthan to caterers, retail shops, restaurants, and canteens.

   Farm Again IoT devices are used to monitor and record moisture content and soil conditions, with pipes for water and fertiliser inputs.

   Farmizen is a mobile-based platform that lets users grow vegetables and fruits on mini-farms, and monitor the process of growing food on a real-time basis. The startup also provides recommendations based on real-time inputs from the field as well as pre-defined schedules for over 50 different types of crops.

Table 11.3: State-wise Numbers of FPCs in India

States	*2017*	*2018 Per cent Increase*	*2019 SFAC Promoted FPOs as on 31.07.2019*
Andhra Pradesh	2	0.19	7
Arunachal Pradesh	13	1.24	2
Assam	37	3.53	12
Bihar	19	1.81	29
Chhattisgarh	18	1.72	26
Delhi	4	0.38	4
Goa	2	0.19	2
Gujarat	34	3.24	21
Haryana	24	2.29	23
Himachal Pradesh	5	0.48	6
Jammu and Kashmir	2	0.19	2
Jharkhand	8	0.76	10
Karnataka	111	10.59	119
Kerala	10	0.95	-
Madhya Pradesh	127	12.12	143
Maharashtra	85	8.11	99
Manipur	4	0.38	6
Meghalaya	3	0.29	3
Mizoram	-	-	1
Nagaland	2	0.19	2
Odisha	33	3.15	41
Punjab	7	0.67	7
Rajasthan	38	3.63	48
Sikkim	29	2.77	30
Tamil Nadu	62	5.92	11
Telangana	54	5.15	20
Tripura	4	0.38	4
Uttar Pradesh	204	19.47	49
Uttarakhand	45	4.29	7
West Bengal	62	5.92	83

Source: SFAC 2019 and Anirban *et al.*, 2018.

3. New Social Initiatives with the Recent Concepts of FPOs/FPCs

FPCs are expected to overcome the constraints of farmers imposed by the small size of their individual farms by leveraging the collective strength and bargaining power to access financial and non-financial inputs, services and technologies (SFAC 2014); enhance incomes, reduce costs of input purchases along with transaction costs, create opportunities for involvement in value-addition including processing, distribution and marketing (Agarwal 2010).

As on March 2019, there were 2,083 FPOs supported by NABARD and 551 supported by SFAC in India. As depicted from the Table 11.1. Madhya Pradesh having highest number of FPCs, followed by Karnataka and Maharashtra, West Bengal and Uttar Pradesh (Table 11.1). IFAD (2004) opined that in rural areas, farmers' organizations (FOs) are the nearest and often only institutions providing essential goods and services to the rural poor and helping them to break out from the poverty cycle. The role of the Farmer Producer Organisations (FPO) is critical in the development of comprehensive and workable supply chains.

4. Digital Initiatives for Better Information Access and Knowledge Management in Agriculture

Agriculture requires generous knowledge transfer to and among farmers, together with information about fruitful agricultural practices, new expertise or controls of pest and disease eruptions, and new markets. ICTs can support directly farmers' access to timely and relevant information, as well as empower the conception and sharing of knowledge of the agricultural community themselves (Kumar *et al.*, 2017).Knowledge is reflected as the fourth production aspect after labour, land and capital (AFAAS, 2011) and is predominantly precarious in the agrarian sector. Agriculture Knowledge Management System is a platform enabling extraction, storage, retrieval, combination, transformation, visualization, investigation, dissemination and application of knowledge. Different kinds of portals can be used in KM applications to enable various users to interact and establish relationships in order to identify, conceptualize and develop, utilize, apply and evaluate the information and knowledge for shared learning (Malekmohammadi, 2009). Different knowledge management initiatives have been produced significant effect on farming communities. Some of the initiatives like Agropedia, Rice Knowledge Management Portal (RKMP), DKMA of ICAR, some of the expert systems like Agridaksh of IASRI and many more has been satisfied the consumers. Various others initiatives like Web Portal- Krishi Vigyan Kendra Knowledge Network Portal, dynamic portal having read and write facility, launched on 8 July 2016 for regular monitoring of KVKs and to provide information and advisories to the farmers. mKisan Portal- 90 lakh farmers are being provided advisory by the KVKs. Farmers' Portal is an endeavour in this direction to create one stop shop for meeting all informational needs relating to Agriculture, Animal Husbandry and Fisheries sectors production, sale/storage of an Indian farmer. With this Indian Farmer will not be required to sift through maze of websites created for specific purposes. Kisan call centre reaching effectively to all sections of farmers.

Some of the ICRISAT initiatives to cater the information needs are:

- ☆ **Krishi Vani:** Voice message–based advisory system.
- ☆ **Krishi Gyan Sagar:** Data-based information system piloted.
- ☆ **ISABELA (Imagery for Smallholders Activating Business Enterprises and Leveraging Agriculture):** Satellite images used to delineate land boundaries accurately.

- Climate information made available to farmers on mobiles
- **Sowing Application:** Essential information relatedto sowing, based on weather data and forecast models, imparted to farmers via mobile app.
- ICT-based agri-start-ups incubated by ICRISAT begin software development to support FPOs nationwide.
- App for pest and disease recognition: PEAT Gmbh'sPlantix mobile app customized to ICRISAT's mandate crops, enabling farmers to identify pests and diseases using their mobile phones and obtain remedial measures.

5. Artificial Intelligence (AI) in Agriculture

NASCCOM predicts that by 2022, a startling 46 per cent of the Indian workforce will be engaged in entirely new jobs that do not exist today or jobs that have radically changed skill sets. India had initiated different programmes to supports these cutting edge technologies and use their importance for the growth of the country *viz.* Digital India, Start Up India, Make in India would help India establish itself as a knowledge and digital society. Recognising the importance of Artificial Intelligence (AI), the National Institution for Transforming India (NITI) Aayog has initiated in establishing the National Program on AI, with a view to guide the research and development in new and emerging technologies, and publish discussion paper on National Strategy for Artificial Intelligence. NITI Aayog has adopted a threepronged approach for building a vibrant AI ecosystem in India and collaborating with various experts and stakeholders, to position itself among leaders on the global AI map with a unique brand of AI for All (#AI for All). **Precision agriculture**

Precision agriculture means application of precise and correct amount of inputs like water, fertilizer, pesticides *etc.* at the right time to the crop, for increasing its productivity and maximizing its yields. Precision agriculture management practices can significantly reduce the amount of agrochemicals and other crop inputs used while boosting yields. Farmers thus obtain a better return on their investment by saving on water, pesticide and fertilizer. Application of right amount of agrochemicals in the right place and at the right time, is proving beneficial for crops, soil and groundwater.

Start Ups in Artificial Intelligence

1. **Blue River technology-weed control:** The ability to control weeds is a top priority for farmers and an ongoing challenge as herbicide resistance becomes more common place. Today, an estimated 250 species of weeds have become resistance to herbicides. This California-based startup company developed a robot called See and Spray which reportedly leverages computer vision to monitor and precisely spray weeds on cotton plants. Precision spraying can help prevent herbicide resistance.
2. **Harvest CROO robotics – crop harvesting:** Automation is also emerging in an effort to help address challenges in the labour force. Harvest CROO Robotics has developed a robot to help strawberry farmers pick and pack their crops. Harvest CROO Robotics claims that its robot can harvest 8 acres in a single day and replace 30 human labourers.

3. **PEAT – machine vision for diagnosing pests/soil defects:** Berlin-based agricultural tech startupPEAT, has developed a deep learning application called Plantix that reportedly identifies potential defects and nutrient deficiencies in soil. Analysis is conducted by software algorithms which correlate particular foliage patterns with certain soil defects, plant pests and diseases. The image recognition app identifies possible defects through images captured by the user's smartphone camera.
4. **Sky Squirrel technologies Inc. – drones and computer vision for crop analysis:** The company aims to help users improve their crop yield and to reduce costs. Sky Squirrel uses algorithms to integrate and analyze the captured images and data to provide a detailed report on the health of the vineyard, specifically the condition of grapevine leaves. Since grapevine leaves are often telltales for grapevine diseases (such as molds and bacteria), reading the "health" of the leaves is often a good proxy for understanding the health of the plants and their fruit as a whole.
5. **aWhere – satellites for weather prediction and crop sustainability:** A Colorado based company uses machine learning algorithms in connection with satellites to predict weather, analyze crop sustainability and evaluate farms for the presence of diseases and pests.

6. Climate Smart Agriculture through Digital Platforms

The impacts of climate change on crop yields indicate that yield losses may be up to 60 per cent, depending on crop, location, and future climate scenario (Challinor *et al.* 2014) Promotion of Climate Smart Agriculture (CSA) through the Climate-Smart Village (CSV) approach for technological and institutional options for dealing with climate change in agriculture. Providing weather-based insurance; Use of ICT for dissemination of climate information based agro-advisories for Development of decision-support tools for planning and investment. Emphasis on organizations as mechanisms for linking between national-level and community-level adaptation, and associated range of activities, Selection of technologies to local contexts, mapping local institutions and working in partnership across institutions, Gender and social inclusion in climate change adaptation and developing Success factors include participative and locally driven vulnerability assessments and tailoring of adaptation measures.

Table 11.4: Climate Smart Technologies for Farming Community

Sl.No.	*Technology*	*Adaptation/Mitigation Potential*
1.	**Water-smart**	Interventions that improve water use efficiency
a)	Rainwater Harvesting (RH)	Collection of rainwater not allowing to run-off and use for agricultural in rainfed/dry areas and other purposes on-site
b)	Drip Irrigation (DI)	Application of water directly to the root zone of crops and minimize water loss
c)	Laser Land Levelling (LL)	Levelling the field ensures uniform distribution of water in the field and reduces water loss (also improves nutrient use efficiency

Sl.No.	Technology	Adaptation/Mitigation Potential
d)	Furrow Irrigated Bed Planting (FIBP)	This method offers more effective control over irrigation and drainage as well as rainwater management during the monsoon(also improves nutrient use efficiency)
e)	Drainage Management (DM)	Removal of excess water (flood) through water control structure
2.	**Energy-smart**	Interventions that improve energy use efficiency
a).	Zero Tillage/Minimum Tillage (ZT/MT)	Reduces amount of energy use in land preparation. In long-run, it also improves water infiltration and organic matter retention into the soil
3.	**Nutrient-smart**	Interventions that improve nutrient use efficiency
a)	Site Specific Integrated Nutrient Management (SINM)	Optimum supply of soil nutrients over time and space matching to the requirements of crops with right product, rate, time and place
b)	Green Manuring (GM)	Cultivation of legumes in a cropping system. This practice improves nitrogen supply and soil quality
c)	Leaf Color Chart (LCC)	Quantify the required amount of nitrogen use based on greenness of crops. Mostly used for split dose application in rice but also applicable for maize and wheat crops to detect nitrogen deficiency
d)	Intercropping with Legumes (ICL)	Cultivation of legumes with other main crops in alternate rows or mixed. This practice improves nitrogen supply and soil quality
4.	**Carbon-smart**	Interventions that reduce GHG emissions
a)	Agro Forestry (AF)	Promote carbon sequestration including sustainable land use management
b)	Concentrate Feeding for Livestock (CF)	Reduces nutrient losses and livestock requires low amount of feed
c)	Fodder Management (FM)	Promote carbon sequestration including sustainable land use management
d)	Integrated Pest Management (IPM)	Reduces use of chemicals
5.	**Weather-smart**	Interventions that provide services related to income security and weather advisories to farmers.
a)	Climate Smart Housing for Livestock (CSH)	Protection of livestock from extreme climatic events (*e.g.* heat/ cold stresses)
b)	Weather based Crop Agro-advisory (CA)	Climate information based value added agro-advisories to the farmer
c)	Crop Insurance (CI)	Crop-specific insurance to compensate income loss due vagaries of weather
6.	**Knowledge-smart**	Use of combination of science and local knowledge
a)	Contingent Crop Planning (CC)	Climatic risk management plan to cope with major weather related contingencies like drought, flood, heat/cold stresses during the crop season
b)	Improved Crop Varieties (ICV)	Crop varieties that are tolerant to drought, flood and heat/cold stresses
c)	Seed and Fodder Banks (SFB)	Conservation of seeds of crops and fodders to manage climatic risks

Adopted from Khatri-Chhetri, *et al.*, 2017.

Conclusion

Now need to bring equilibrium between the demand and supply. We need to focus on markets. We would like to liberalize agriculture, make it more market friendly, so that it is the demand and the prices that become the incentive for the farmers to produce what the global markets demand or the domestic markets demand. Digitization may also be an opportunity for democratization of knowledge, and agroecology is a knowledge intensive system in which information and data should be specific to the local context. In process of achieving the transformation there is need to address the Strategies and approaches for sustainable agricultural development through Digital approaches for secondary agriculture (organic farming) and adoption of Smart approaches for technology Assessment and dissemination. A good Convergence models need to be developed for economic prosperity. Technological Interventions for Climate Smart Agriculture Initiatives for Food and Nutritional Security for Rural Communities should be taken on priorities in terms of farming systems for nutrition.

REFERENCES

AFAAS. 2011. Concept and learning framework for the African Forum for Agricultural Advisory Services (AFAAS). AFAAS, Kampala, Uganda and FARA, Accra, Ghana.

Agarwal, B. 2010. Rethinking agricultural production collectivities. *Economic and Political Weekly*, 45(9):64-78.

Alexandratos, N., and J. Bruinsma. 2012. World agriculture towards 2030/2050: the 2012 revision. ESA Working paper No. 12-03. FAO, Rome, Italy.

Bachaspati, S. and Pathak, H. 2018. Impact of e-NAM on arrival and prices of major commodities: A case study of Bhatapara APMC in Chhattisgarh. *Journal of Pharmacognosy and Phytochemistry*, 7(4): 3301-3304.

Challinor, A.J., Watson, J., Lobell, D.B., Howden, S.M., Smith, D.R. and Chhetri, N. 2014. A meta-analysis of crop yield under climate change and adaptation. *Nature Climate Change*, 4:287–291.

Dev, M. 2017. Small Farmers in India: Challenges and Opportunities. [*Retrieved from IGIDR Working Paper: http://www.igidr.ac.in/pdf/publication/WP-2012-014.pdf*].

Food and Agriculture Organization (FAO). 2015. The state of food insecurity in the world 2015. Strengthening the enabling environment for food security and nutrition. FAO, Rome, Italy.

Government of India. 2016. State of Indian agriculture, 2015-16. [*Retrieved from http://eands.dacnet.nic.in/PDF/State_of_Indian_Agriculture,2015-16.pdf*].

IFAD. (2004). IFAD Recommendations for eliminating rural poverty and achieving food security, 36th World Farmers' Congress, 29th May- 4th June 2004, Washington DC.

Khatri-Chhetri, A., Aggarwal, P. K., Joshi, P. K., and Vyas, S. 2017. Farmers' prioritization of climate-smart agriculture (CSA) technologies. *Agricultural systems*, 151: 184-191.

Kumar, S., Sangeetha, V., Singh, P., Burman, R.R., Bhowmik, A., and Kumar, S.A. (2017). Constraints Faced by Farmers in Utilizing Rice Related Information through Rice Knowledge Management Portal (RKMP). *Indian Journal of Extension Education,* 53(1), *84-89.*

Malekmohammadi, I. (2009). Interpretive Perspective of Knowledge Management Stance in Agricultural Knowledge Information System to Fostering Research/ Extension Linkage. *American Journal of Agricultural and Biological Sciences* 4(3): 230-241.

Meera, S.N.; Jhamtani, A. and Rao, D.U.M. 2004. Information and communication technology in agricultural development: A comparative analysis of three projects from India. *Agril. Res. and Ext. Network.* Network Paper No. 135, pp.13.

MoA and FW. 2011. All India report on number and area of operational holdings. Agriculture Census New Delhi: Ministry of Agriculture, GoI, New Delhi. (http://agcensus.nic.in/document/agcensus2010/agcen2010rep.htm).

MoA and FW. 2016. Agricultural statistics at a glance. Ministry of Agriculture and Farmers Welfare, GoI, New Delhi.

Mukherjee, A., Bahal, R., Burman, R.R., and Dubey, S.K. 2012. Conceptual Convergence of Pluralistic Extension at Aligarh District of Uttar Pradesh. *Journal of Community Mobilization and Sustainable Development.* 7(1 and 2): 85-94.

Mukherjee, A., Singh, P., Ray, M., Satyapriya, and Burman, R. R. 2018. Enhancing farmers income through farmers' producers companies in India: Status and roadmap. *Indian Journal of Agricultural Sciences,* 88(8): 1151-1161.

NABARD. 2018. All India rural financial inclusion survey 2016-17. National Bank for Agriculture and Rural Development, Mumbai.

Reddy, A. 2018. Electronic national agricultural markets: the way forward. *Current Science,* 115(5): 826

SFAC. (2014). Krishi sutra 2: Success stories of farmer producer organisations. Small farmers agribusiness consortium, New Delhi. [*Retrieved from http://sfacindia.com/PDFs/Krishi-Sutra(Version2).pdf*].

Williamson, O.E. 1985. *The Economic Institutions of Capitalism,* Free Press, New York, USA.

Yadav, K.; Rasheed Sulaiman, V.; Yaduraju, N.; Balaji, V. and Prabhakar, T. 2015. ICTs in knowledge management: the case of the Agropedia platform for Indian agriculture. *Knowledge Mgt. for Dev. J.,* 11(2).

Chapter 12

Mobile Based Agro-advisory in Meghalaya: Lesson Learnt in First Phase

Ram Singh[1], *M. Premjit Singh*[2], *R.K. Singh*[3] *and J.K. Chauhan*[4]

[1]*Professor,* [3]*Asstt. Professor (Extension),* [4]*Professor (Extension), CPGS-AS, CAU (Imphal), Umiam – 793 103, Meghalaya*
[2]*Hon'ble Vice-Chancellor, CAU, Imphal, Manipur*

The ICT is a partial tool to enhance the productivity and profitability of the farmer. Various initiatives in IT (Information Technology) were launched in the state to boost up the yield of crops in order to enhance the livelihood and income of the farmers. Among the initiatives one agro-advisory laboratory has been established under the programme of "Development and Deployment of Mobile based Agro- advisory system in North-East India (hereafter m4agriNEI)" in College of Post Graduate Studies, Central Agricultural University, Barapani. The present study was conducted in 11 villages of Ri-Bhoi district of Meghalaya. Over 218 farmers were randomly selected from the eleven villages. Data were collected by using pre-tested structured interview schedule through personal interview method. The tabular analysis was done to study the perception and accrual of benefits of the programme for the farmers located in remote villages. The farmers expressed that the information gathered through agro-advisory were very useful and helpful for them. The programme has gained lot popularity among the farmers of Meghalaya and the sufficient numbers of beneficiaries were found to be benefitted by using the agro-advisory in the crop as well as in livestock sector. Among the crops, the ginger crop being a cash crop of the state need more care right from its planting to harvesting. The agro-advisory on complete package and practices especially in selection of rhizome, its treatment, maintaining spacing during the planting and also the tips of proper care during harvesting provided by experts of agro-advisory really became the boon in production of ginger in the state. Providing the artificial insemination and vaccination in pig and piglets encourage the farmers to come forward to access more and more information from Agro-advisory services. Keeping the success in view the funding agency has extended the finance support two times. Hence, to provide the agro-advisory in sustainable manner, the Government of Meghalaya taken over the Laboratory which has added one more component viz, mobile marketing vans for agri-produce from farmers to market in the state

Keywords: *Lesson, Meghalaya, Mobile, Agro-advisory, Tribal.*

Background

Meghalaya state is the homeland of tribes namely *Khasi*, *Jaintia* and *Garo*. The schedule tribes comprise more than 86 per cent of the total population in the state. The tribal people of Meghalaya are the world's largest surviving matrilineal culture. They follow the traditional matrilineal norm. The state is basically rural based with agriculture as the predominant role in the state's economy around 80 per cent of its population depending entirely on agriculture for their livelihood (GOM, 2015). The cropping intensity of the state is estimated as 119 per cent. It is low compared to national level. The state's agricultural system is predominantly traditional. The state being agrarian, economy also depends on animal husbandry. The various big animals such as cattle (767 thousand), Buffaloes (18 thousands), pigs (420 thousand) small animals (2821 thousand). Fish production annually accounted as 3.9 thousand ton in the state (Anonymous, 2008). According to basic statistics of NER, 2006, the overall geographical land-to-man ratio for the NE region (0.67 ha/Person) is much higher than the national average (0.32 ha/Person). The soil and climatic condition of the state is suitable for growing different types of agricultural crops from cereals to fruits in both tropical and temperate climatic environment occurring on different altitudes. Though, 81 per cent of the population depends on agriculture, the net cropped area is only about 9.87 per cent of the total geographical area of the state (GOM, 2015).

Food grains are the most important crops in the state and cover an area of over 60 per cent of the total crop area. The state is still deficit in food grains by 1.22 lakh tonnes annually to feed a population of 2.3 million. This is due to a lot of constraints, such as the undulating topography, transport and communication problem, population dispersal pattern, inadequate credit support, poor marketing system, *etc.* State's climate is very much favourable to produce all types of crops such as cereals (rice and maize), fruits (mandarin and pineapple), spices (ginger and turmeric), medicinal crops, flowers and cash crops (cashew nut and tea leaves) *etc.* Mostly, the farmers in rural areas have to deal with frequent crop failure and animal illness due to limited accessibility of agro-advisory particularly farmers of remotely located villages in difficult terrain. They have to walk for long distance to get solutions of their problems and it incurs time and money a lot. As everything else wait, but agriculture never (Nehru, 1947) becoming more relevant for such type of area where getting information/advisory and same time providing the advisory to the farmers located in remote village is really a challengeable task.

About the Agro-advisory System

Agriculture in the state could benefit tremendously with the application of ICTs (Information and Communication Technologies) especially in bringing changes to socio-economic conditions of the poor in the backward areas. Therefore, different initiatives in IT sector were launched in the state to provide the agro-advisory to farmers. Among the initiatives an agro-advisory system launched under the project of "Development and Deployment of Mobile based Agro- advisory system in North-East India (hereafter m4agriNEI)" by the Central Agricultural University (hereafter CAU) at Barapani in the state. The system is started with collaboration of Media Lab Asia, New Delhi funded by IT department, government of India, New Delhi. The m4agriNEI is an integrated system with a combination of Web, Interactive Voice Response System (IVRS) and mobile technologies for dissemination of farm and farmer specific advices/information at user desired mode and time it has Toll Free Number (1800-345-3700). It is a mobile based pull and push system where agriculture related information can be pulled/pushed by the farmers using their mobile phones. There is a mobile interface at front end for the farmers and web interface at the back end for the agri-experts. The system allows transmitting the data through voice, texts, images and videos from both end (farmers to expert and vice versa). Also, the farmer can call the system to get any information as well as to get agro advisory. Farmer receive information (SMS/Voice Call/Data on Smart Phone) for only those services for which he has subscribed and has an option at a later date to either select some more services or unsubscribe to some of the existing services. The system is connected to a centralized database, which has all information of farm, farmer and previous transactions. The experts at back end (data centre and virtual expert) can access to the database of the farmers while responding the farmer's queries and only registered user can access the m4agriNEI system. There is a mobile interface at front end for the farmers and web interface at the back end for the agricultural experts (Level-I and Level-II). Further, designated Farmer Co-ordinators and rural youth facilitate the registered farmers in getting farm information and knowledge and also they provide feedback to the m4agriNEI system. The system is connected to a centralized database, which have informations on farmer, farm history, and previous interactions (Figure 12.1).

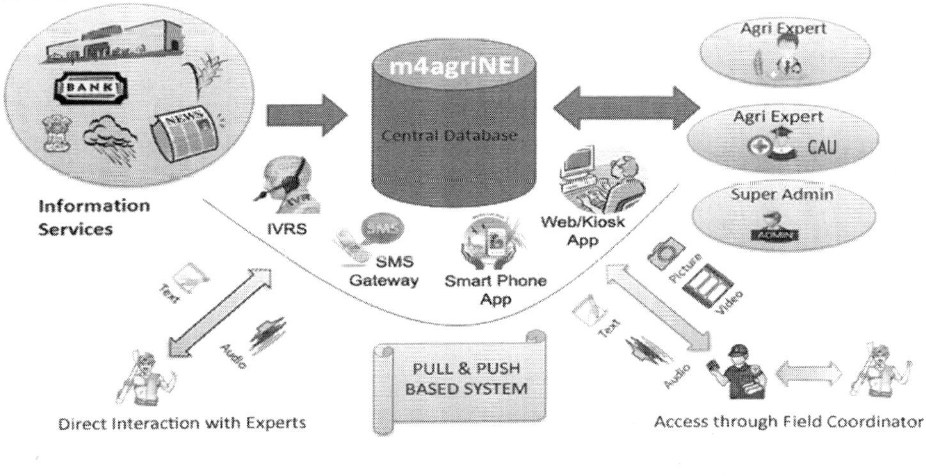

Figure 12.1: System Architecture of Agro-advisory and Activity Framework.

Components of Agro Advisory

Agro-advisory (m4agriNEI) is an integrated system with a combination of Web, IVRS and Mobile applications for dissemination of farm and farmer specific advises/information at user desired mode and time (Figure 12.2). Each component is described as under:

Mobile Application for Smart Phones (Android and Windows)

- Standalone Application.
- Offline query aggregation capabilities.
- Data Synchronization at hot spots/areas of data connectivity (store and forward).
- Offline capability of display of last synchronized data.
- Coordinator information system (profile page, visit scheduling, list of registered farmers).
- Farmer registration and profile.

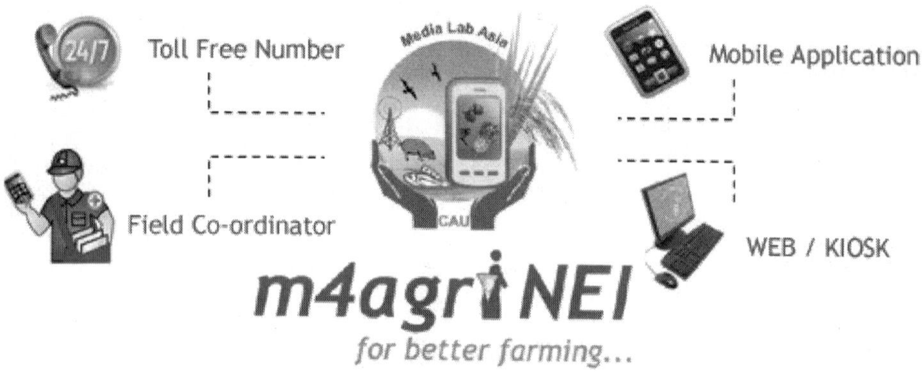

Figure 12.2: Component of Agro Advisory.

Web Based Application

- Colour coded iconic based logins for various type of users.
- Expert support system linked with authentic content/information service providers.
- Information dissemination and aggregation system (multimodal).
- Centralized common database for web, mobile and IVRS applications.
- Currently available in English, Khasi and Garo.
- Reports and analytics.

Interactive Voice Response System (IVRS) Based Application

- Call incoming facility on expert's computer.

- ☆ Call forwarding and recording facility in case if the experts are not available.
- ☆ Intelligent enough to route the call to the relevant experts.
- ☆ 24x7 query registration facility for farmers.

Methodology

The agro-advisory service is being provided to 2813 numbers of farmers of 47 villages of Ri-bhoi district. The registered farmers on a specific form are being served by the mobile based agro-advisory service system. The system displaying the all history of the farmers which include the information of name, village name, photo of the farmer and crops grown by the farmers as and when call landing to system. An impact assessment of the project was conducted in 11 villages of Ri-Bhoi district of Meghalaya. A sample of 218 farmers has been drawn randomly from the eleven selected villages. Data were collected by using pre-tested well structured interview schedule through personal interview method. The simple tabular analysis was done to study the impact of the programme in the state.

Results and Discussion

Analysis of Queries

The query analysis is presented in Table 12.1 and Figure 12.3. The farmers raised query about the seed and its source which was accounted of 32.91 per cent of the total queries. Improved seeds of ginger, tomato, paddy, French bean, maize, chilli, flower and other vegetables were the seeds wanted by farmers. Piggery piglets, poultry chicks (Kuroiler, layers) and fish fingerlings were queries among livestock which enquired by the farmers. The second highest number of queries raised by the farmers was on crop disease and pest management and accounted to be 17.08 per cent of the total queries. Farmers enquired on modern scientific preventive measures of various diseases and pests. Crop disease and pest management of ginger (soft rot disease, shoot borer), preventive measures of paddy diseases (*blast disease, brown spot, bacterial blight, Udbatta* disease) and pests (*Gundhi bug, cutworm, white grub, rice leaf folder, rice case worm, and rodents*) and for french bean (*leaf miners and aphids*). The 14.90 per cent of query rose to know about the agro-advisory system and toll free number. Similarly, 12.36 per cent of the queries were raised about livestock management, in which maximum on piggery which related to skin infection, de-worming of pig, artificial insemination, vitamins, *coughing*, diarrhoea, swine fever, castration which was followed by preventive measures of *rani khet coughing, diarrhoea* disease of poultry. Further, the query on crop cultivation and improvement of crops was observed to be 6.96 per cent of total queries in which information enquired on scientific management and improvement of cultivation of different crops like ginger, onion, paddy, tomato, cauliflower, orange, mustard and other crops. An about 6.87 per cent of the queries were raised about the welfare schemes launched by state government. Some of the farmers were found to be very much interested on trainings of which 5.21 per cent of the queries were on trainings related to piggery, poultry, ginger cultivation, fish farming, mushroom cultivation, paddy cultivation, beekeeping, flower cultivation, vegetable cultivation *etc.* A few queries were raised on nutrient management, fishery management, marketing *etc.*

Table 12.1: Query Analysis

Type of Query	*Total*	*Percentage (per cent)*
Source of seeds	1073	32.91
Crop disease and pest management	557	17.08
Enquired about the project	486	14.9
Livestock management	403	12.36
Crop cultivation and improvement	227	6.96
Scheme	224	6.87
Training	170	5.21
Other	60	1.84
Marketing	34	1.04
Fishery management	12	0.36
Nutrient management	14	0.42
Total	**3260**	**100**

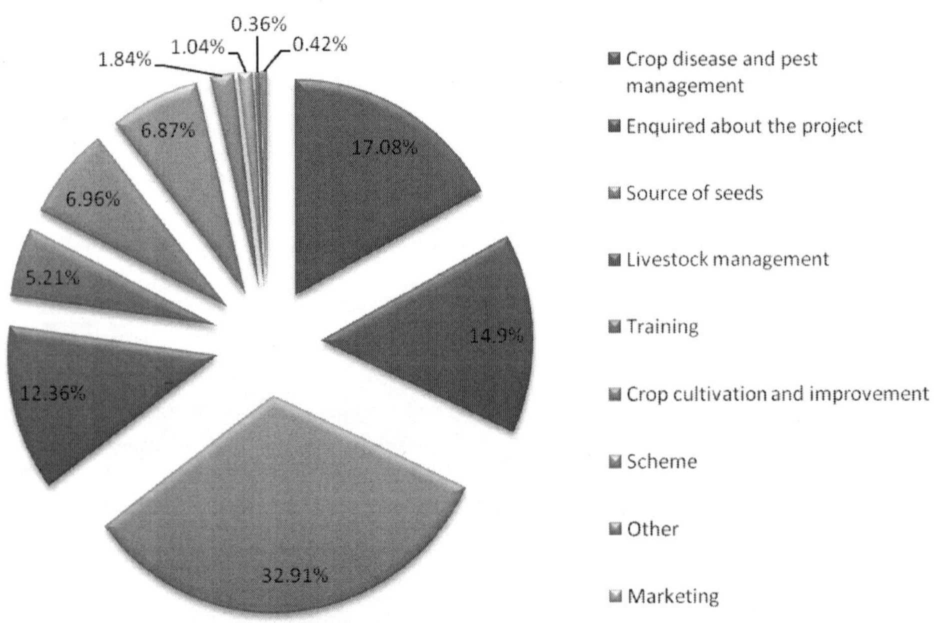

Figure 12.3: Analysis of Queries

Monthly Analysis of queries

It can be observed from the Figure 12.4 that the queries raised by the farmers in the beginning (June'13) accounted to be only 0.42 per cent of the total queries. This may be due to unawareness of farmers about toll free number and due to lack of initial inhibition of the farmers to make calls to the agro-advisory But as the service moved forward, the farmers have started to raise queries through the toll free number of which the highest percentage of the queries was observed in the month of May 2014 and it was estimated to be 8.34 per cent and it may due to the fact that the May month is an active growing period for rice, ginger and turmeric crop in the state.

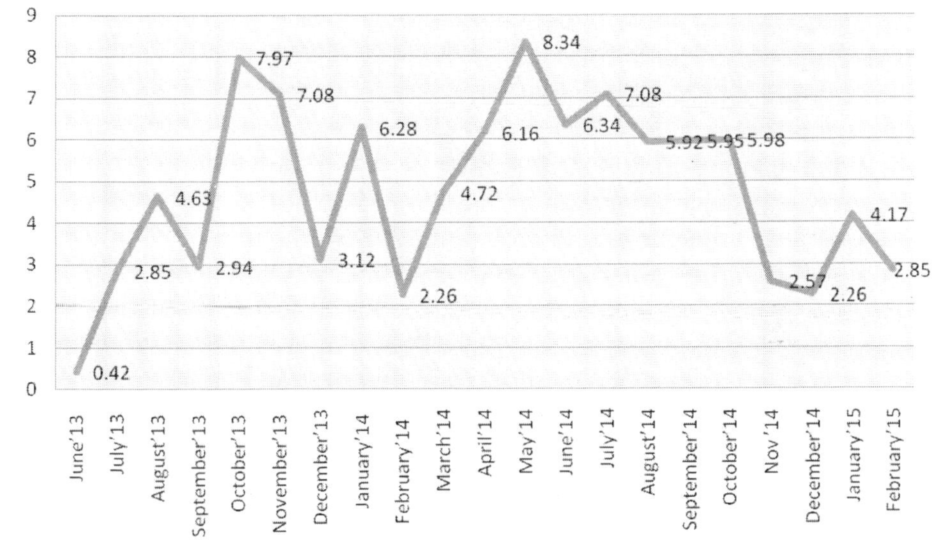

Figure 12.4: Monthly Analysis of Queries.

Queries from different Channel

The different channel used to rise the by the farmers is presented in Table 12.2 and Figure 12.5. Most of the farmers are reluctant to raise queries by themselves through the toll free number so overcome to this problem assistance of Farmer Coordinator were asked to assist the farmers to raise query to the agro-advisory lab. But after much effort made by the project team through awareness, training programmes conducted, field visits by level-1 and level-2 experts the farmers started to raise queries on their own. Farmers raised their queries through different channel of which 77.20 per cent of the queries were raised with the help of the Farmer

Table 12.2: Queries Raised by the Farmers through different Channel

Channel	No. of Queries	Percentage (per cent)
Farmer coordinator (FC)	2517	77.20
Village level entrepreneur (VLE)	26	0.79
Level I expert	548	16.80
Own call from farmers	169	5.18
Total	**3260**	**100**

Coordinator and followed by Level-I expert assistance which was accounted to be 16.80 per cent. An around 5.18 per cent of the queries were raised by farmer themselves without any assistance through toll free number.

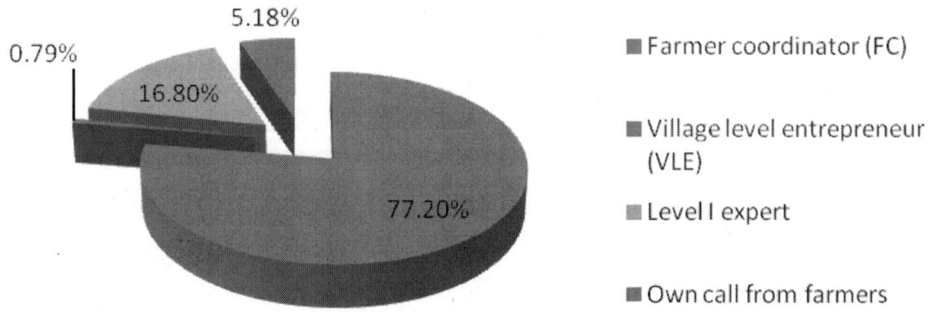

Figure 12.5: Queries from different Channel

Perception about the Mobile Based Agro-Advisory

As a whole, the project design was flexible enough to accommodate changing situations of farming system of the farmers. All of the respondents were fully aware of the project through different sources: 52.29 per cent through repeated on-site awareness/training programmes, 37.61 per cent through farmer coordinators and 10.09 per cent through interaction with the project team members during their field visit programmes (Table 12.3).

Table 12.3: Awareness About the Mobile Based Agro-Advisory (n=218)

Source	*Response*
Awareness/Training programmes	114 (52.29)
Farmer Coordinator	82 (37.61)
Through project team home/field visit	22 (10.09)

Figures in parenthesis indicate percentage to total.

The information provided through the m4agriNEI team was clear to 92.20 per cent, easily understandable to 91.28 per cent, timely to 76.14 per cent, complete to 41.73 per cent and practicable/adaptable in field conditions to 37.61 per cent (Table 12.4).

Table 12.4: Perception about the Quality of Information and Advisory Support (Percent)

Perception	*Response*
Clear information.	92.2
Easily understandable.	91.28
Timely information.	76.14
Complete.	41.73
Practicable/adaptable in the field conditions.	37.61

Majority of the respondents were satisfied with the information provided by the m4agriNEI team on pest and disease management on ginger and 82.11 per cent of respondents responded in this favour. About 69.66 per cent of participants reported that training provided on ginger cultivation was beneficial for them. Around 19.26 of farmers were satisfied with the training conducted on improved variety of ginger. Hence, training and information provided by the agro-advisory to the farmers on ginger was appreciated by majority of the farmers since, the ginger being a high value crop is contributing lion share in the livelihood and income of the farmer. Other hand 66.05 per cent of farmers reported that advisory provided on piggery was beneficial. The advisory given on rice crop only 7.33 per cent of rice growers reported it as beneficial since rice is grown long back and mostly of farmers are very adamant to cultivate rice in traditional method and another reason the rice is grown only for home consumption not for commercial purpose (Table 12.5).

Table 12.5: Knowledge Gained through Mobile Based Agro-Advisory (Per cent)

Sectors	*Response*
Ginger	
Pest and disease management	82.11
Training	69.66
Improved variety	19.26
Piggery	66.05
Rice	7.33

The information provided through m4agriNEI service was fully utilized by 17.88 per cent of the respondents in terms of piggery while 26.60 per cent have partially utilized that information. In terms of information related to ginger crop, 2.75 per cent of the respondents have fully utilized the information while 33.94 per cent have partially utilized that information. In case of rice, 2.75 per cent of the respondents have partially utilized the information (Table 12.6).

Table 12.6: Extent of Information Utilization by the Farmers Provided through Mobile Based Agro-Advisory (Per cent)

Sectors	*Extent of utilization*	
	Full	*Partial*
Ginger	2.75	33.94
Piggery	17.88	26.6
Rice		2.75

About 19.72 per cent of the respondents expressed that they were benefitted due to m4agriNEI service with regard to ginger cultivation technology and sources of availability of healthy rhizomes (Table 12.7).

Table 12.7: Perceived Accrual of Benefits by the Farmers from Agro-Advisory Services (Per cent)

Sectors	*Response*
Animal husbandry	19.72
Progress in crops	18.80

Conclusion

The farmers expressed that the information gathered through agro-advisory were very useful and helpful for them for collaborative learning with peers and experts. The farmers found to be taking interest to access the information in management of crops and animal husbandry. The seed for crops and pig management were the major aspects on which farmers were interested to get information. At initial stage of the programme, the sufficient numbers of beneficiaries were found to be benefitted by using the agro-advisory. Moreover, its success has provided and extended financial support twice and it has attracted the state Government of Meghalaya for taken over under the administration and finance control of local government. It led the sustainability forever the intervention. Hence, to provide the agro-advisory in sustainable manner, the convergence of such type of programmes with state department in other states of the region considering the technical help from Central Agricultural University along with KVKs personnel is highly recommended so that it will help to uplift the livelihood of rural tribal in rapid way.

Acknowledgment

The authors are highly grateful to funding agency of DeitY, MITC, GoI, New Delhi to provide the funds to initiate the programme entitled "Development and Deployment of Mobile based Agro-advisory system in North-East India (m4agriNEI)" from which this paper has been prepared. The authors are also thankful to competent authority of esteemed Central Agricultural University, Imphal, Manipur to execute this project.

REFERENCES

Anonymous. 2008. Investment potential in North Eastern India, prepared by BIRD and IMRB for Royal Thai Embassy, New Delhi-India

Basic statistics of NER. 2006. Govt. of India, Nongrim Hills, Shillong

GOM. 2015. Meghalaya Agriculture Profile, Shillong, Meghalaya. [*Retrieved from http://meghalaya.gov.in:8080/megportal/stateprofile*].

Nehru, J.L. 1947 in context of the Bengal Famine of 1942-43 and the acute food scarcity prevailing in the country in 1947.

Chapter 13

Multisided, Multitier Extension Approaches and Digital Strategies of Veterinary Varsity for Sustainable Livestock Development

H.K. Verma and Simarjeet Kaur

Directorate of Extension Education,
GADVASU, Ludhiana, Punjab

Livestock farming is undertaken on traditional and orthodox ways by majority of farmers in India, which are either small, marginal or landless. They possess majority of the livestock and lack scientific knowledge for rearing them. Govt. of India is stressing on doubling of farmers income which is only possible through livestock farming either on commercial or through integrated farming. Punjab is known for its Agriculture pursuits. Around seventy percent people of the state are involved in agriculture and allied activities and more than 83 per cent of the total land in the state is under agriculture. In 2016-17, agriculture had 23.96 per cent share in State's GDP while livestock had 8.63 percent share (Statistical Abstract Punjab, 2017). In the recent past, the share of livestock GDP in agriculture increased steadily. Livestock economy including fisheries contributed 29.60 per cent of the agricultural GDP in 2000-01 while the share increased to 36.01 per cent in 2016-17 and 37.65 (9.10 per cent) in 2017-18. Livestock sub-sector has registered highest growth among all the sub-sectors of agriculture in India while Punjab is rated among the highest for the share of livestock output to agricultural output (Jha *et al.*, 2009). Thus, livestock sector proved to be an important growth engine for the overall development of agricultural sector in the state. 'Skill development- Skilled India' programmes give further impetus to livestock farming.

Livestock productivity of Punjab including fresh water fisheries is highest in the country, however the major constraints faced by livestock sector are small herd size or livestock units, poor extension services concerning livestock sector, lack of proper and scientific nutrition interventions for different categories of animals, insufficient prophylaxis, poor value addition of the produce procured from small and marginal farmers, absence of culling policy for stray animals and male calves, poor linkages with consumer market *etc.* However, few farmers have adopted direct marketing and few self help groups or producer organizations are operational, but not sufficient to take full harvesting of the livestock productivity, later being in the hands of middlemen. Directorate of Extension Education, Guru Angad Dev Veterinary and Animal Sciences University, Ludhiana has adopted a farmer friendly cum multisided, multitier and all-round approach to make livestock farmers acquainted with latest trends of farming. Innovation and research outcome of the university over the years have led to the propagation of various livestock development strategies to the farmers, through various extension methodologies. The main technologies transferred so far are mainly in the field of breeding, nutrition, reproduction, and health management, value addition of produce, newer procedures and methods for productivity enhancement.

> *.......Extension is a service and way of life. It is a commitment, one has to come to the level of farmers for infusing information into farmers ' mind so, that adoption is full for enhancing livestock productivity as per societal needs. Extension cannot be successful through half hearted approach cum happens attitude, however it has to be to make it happen attitude.......*

In India, there are huge economic losses in animals occurring due to diseases amounting to more than one lac crore annually. These losses can be minimized by pinpoint diagnosis through farmer's education. There exists information burst at the institutions but the ground reality is that the percolation to the field level is poor (only 5 per cent farmers have access to animal related information compared to 40 per cent in crop and agriculture, NSSO). So, information crunch livestock farmers need information for getting higher as well as healthy productivity. It makes the way for Extension to come into play for penetration of information to them. This necessitates strong, aggressive and focused approach in field which can be in tandem with institutions as they bring the nascent procedures and their methodologies for diffusion. The dissemination of knowledge and technology to the farmers/livestock owners is a regular feature for meeting one of the basic mandates of the university. The fulfilling the three mandates of teaching, research and extension has lead to grab the ***top position*** among the state veterinary varsities of India (ICAR ranking, 2019).

GADVASU's Institutes/Units Involved in Extension Activities

1. College of Veterinary Science
2. College of Dairy Science and Technology
3. College of Fisheries
4. Veterinary Polytechnic

5. Regional Research and training Centres- Bathinda, Hoshiarpur and Tarn Taran.
6. Krishi Vigyan Kendras (KVKs)- Barnala, Tarn Taran and Mohali
7. Skill development centre
8. Farmers' information centre

Major Extension Activities

1. **Field Camps:** Extension wing organizes animal welfare camps and animal welfare days – in the field for treatment fo sick animals, advice as well as question answers session. On these occasions, demonstrations regarding the collection, dispatch and transport of clinical material like blood, mucous discharge and faeces from the animals were carried out and farmers were made aware of correct method of milking, teat dip, heat detection, acaricide drug application, silage making, computation of ration, vaccination and deworming regimes.
2. **Field Service:** Ambulatory Clinical Service provides treatment facilities and counseling services to the farmers particularly in adopted villages. This facility helps the farmers to get rationale treatment and is a great experience for students to work under field conditions for the whole week.
3. **Trainings:** University conducts various training programmes in dairy farming, poultry farming, pig farming, goat farming, fish farming and value addition of milk, meat, egg *etc.* These trainings help farmers in gaining technical skills and the knowledge to start these activities for self-employment in line with the government's policy of diversification of agriculture. University is also undertaking commercial dairy farming and AI training programmes. Various tailor made courses are also organized by the university. Few out of box training courses like training of butchers for hygienic meat production, guiding milk vendors and sweet manufacturers and training to the feed manufacturers of Punjab have been organized for making available wholesome livestock produce for healthy society. Till now, more than 30 trainings have been organized and around 1500 feed manufacturers got trained.
4. **Farmers fair:** Pashu Palan Mela is organized twice (March and September) a year by GADVASU at its Ludhiana campus. During Pashu Palan Mela, the whole university is showcasing its activities, research and procedures through display inform of exhibitions, demonstrations, lectures and interactions. Chief Minister Award are conferred to innovative and progressive rural farmers in the field of dairy, piggery, goatry, poultry, fish farming and value addition of livestock produce. These awards are granted for bringing healthy competitions among the farming community to perform better. Expert talks are given on the occasion of "Pashu Palan Melas", "Kisan Melas", Field days and 'Regional Kisan Melas', Farmers' day celebrations' *etc.* other regional research stations and KVKs of the University. In addition, the farmers seek advice through personal contact with the experts at these fairs.

5. **Farmers Associations:** Various farmers consortia like PPFA (Progressive Pig farmers association), IFFA (innovative Fish farmers association) Goat farmers associations and two Poultry farmers associations are working under guidance of GADVASU for horizontal learning. In the monthly meeting of these associations, technical lectures from experts are delivered and queries of farmers are also answered. They participate in gala events-farmers fairs of university for sensitizing young students and motivating farmers for diversification.
6. **Skill development centre:** in Animal Husbandry is operational at GADVASU which is affiliated with Agriculture skill council of India (ASCI) and has nine approved qualification packs. Regular trainings are the feature of SDC. Varsity gets request to train farmers, field officers not from Punjab but from all parts of country.
7. **Mass media:** Use of Mass media in the form of Press, Radio, T.V. talks and through Cable channels and Private channels by the University experts, is helping in the dissemination of knowledge in a significant way due to its vast reach. Important aspects of animal health and production are covered from time to time to keep the masses abreast with the latest recommendations. The messages of urgent concern which need immediate attention are flashed through T.V. and radio through DD Jalandhar, Pvt and Cable channels and AIR, Jalandhar as well as FM and gold channels in the local state and North India.
8. **Farmers queries:** The farmers' queries are replied through postal letters, electronic mails and on telephone. The farmers ask various queries on various aspects which are taken care by telephone helpline No. **0161-2414026**. Farmers have direct access to the experts through cell numbers and contact directly.
9. **Messaging:** University has registered farmers (n=>30,000) for sending them regular message through central portal and advisory also sent through whatsapp and text message.
10. **Farmer Information centre**: To provide the farmers the needed information under one roof, "Farmer Information Centre" is operational. This centre helps the visiting farmers, students, unemployed youth, women farmers to get the information related to trainings, education, camps or various inputs as single window system.
11. **Literature** : University has developed huge literature in Punjabi, English as well as in Hindi language for the livestock farmers covering all aspects of livestock, poultry and fish farming. GADVASU'S Punjabi publication, a monthly magazine "Vigyanak Pashu Palan" carries informative articles dealing with Veterinary and Animal sciences. Such published material make a long lasting impression on its readers and can also be used as reference source by veterinarians, field officers as well as by farmers.
12. **Handy tools:** GADVASU has developed management calendars (calf rearing, feeding, general management, vaccination, de-worming and

reproduction) to provide farmers readymade and handy information for up-keeping their animals in a nice way for getting healthy produce for society. Livestock calendar contains information on animal husbandry practices for the whole year.

13. **Linkages**: University has liaison with line departments of Punjab State viz; Animal Husbandry, Dairy Development and Fisheries for undertaking animal welfare activities. These departments organize various animal welfare camps and field days. University provides latest technical know-how in these activities which is a two way learning for field officers as well as budding veterinarians.
14. **Farmer field schools:** Farmers field schools concept has been introduced for better penetration into field through learing of farmers through horizontal approach as farmers learn better from fellow farmers. The schools are operational under KVK, RRTC and department of Veterinary Extension.
15. **Industry:** The faculty members deliver extension lectures to the farmers in collaboration with the other animal welfare agencies of the state like Nestle India, Smith Kline Beecham, Punjab and Sind Bank, Punjab National banks, Markfed, Milfed, IFFCO,various NGO's, Pharmaceuticals and feed companies. This is also unique opportunity for teachers and students to be field ready and to sense the industry pulse for meeting society demands.
16. **Hospital services:** University extends diagnostic facilities constituting biochemical, pathological and parasitological, culture sensitivity, mastitis and rabies testing and post mortem facilities *etc.* the post mortem is done even on holidays and Saturdays and Sundays.
17. **Fire brigade Veterinary Services:** Emergency services when required by the livestock farmers in case of disease outbreaks, accidents, calamities are undertaken without any delay as a sort of fire brigade type service and farmers can contact at no. **0161-2414030.**
18. **Outreach stations:** To bridge the gap between the universities and the field and to know the field and grass root level scenario through Regional/Outreach Stations and three Krishi Vigyan Kendras. These centers are demonstrating developed technologies and research in their operational areas and also organizing various training programmes for unemployed youth/women in Dairy, Poultry, Piggery, Goat farming and also demand driven courses. They are catering to the routine needs of farmers and livestock farmers of the areas along with showcasing of various correct manage mental practices as well dedicated programmes for better productivity.

Digital Strategies

Nobody on this earth whether homosapien or animal is untouched with ICT in the present era and it is proving as a lifeline for all. Now it is very much clear that through digitalization or ICT one can reach to the unreached in far flung rural areas. Fortunately, new developments in information and communication technology and wide spread of Internet access and mobile communications appear to offer some solutions to meet the information and advisory need of small-scale livestock farmers. Access to Internet based information centers is growing rapidly and village information kiosks with user-friendly features such as touch-screens appear to be a promising development. The ICT related initiatives in livestock extension, give extension workers an opportunity to produce customized extension materials that meet the real needs of the farmers they serve. ICT tools are proving handy in filling this knowledge and accessibility gap and extending digital literacy to the end users. As in this century, majority of farmers (including family members) are having access to smart phones along with internet and they can seek desired information from various sources. Many ICT initiatives in livestock sector have been taken by public and private agencies but the scope of improvement still lies. The digitalized services have major advantages as -Access to huge store-house of information is free, Information is available instantaneously 24X7X365, Communication can also be interactive and Information is available from any point on the globe.

> *....Most powerful digitalized tool is the cell phone which has replaced many gadgets and the number of users are around one billion in India. This is the only technology which has brought literacy and alertness even in aged, senile and lethargic people though its user friendly multiple uses. It is a wonderful tool to reach masses through a click of button and has proven useful in emergency situations....*

Major Digitalized Strategies of GADVASU are

1. Mobile advisory
2. Mass Media- Radio, TV, newspaper, Cable network, Pvt. Channels.
3. Messages- Flash, Short and Advisory
4. Videos CD and Audio Cassette-Repeat Breeding, Anoestrus, Abortion, Calf Management
5. GADVASU friends, Farmers groups/Associations,
6. Farmers fairs, Exhibitions, Charts, Models, Live Animals, Specimens,
7. Forums – Workshops, Meetings, Interactive sessions
8. Website(www.gadvasu.in), E- mail
9. Video conferencing
10. Applications- LH and AH (Mobile apps)-
    a. Precision dairy farming App
    b. Dairy Prajanan Mobile App
    c. Goat farming
    d. Pig farming mobile Apps

Inputs-cum-Services Available to Farmers for Livestock Productivity Enhancement

- ☆ 24x7 treatment facilities through Veterinary hospital, equipped with all modern facilities' known as PGI for animals.

- ☆ Surgical operations of conditions like Fracture Impaction, Diaphragmatic Hernia, Tumors, Teat and Urine Blockage and spaying of bitches.
- ☆ Specialized treatment for equine and bovine for lameness and hoof trimming.
- ☆ Treatment of all species of animals like Large, Small, Pet, Bird, Lab, Wild and Zoo animals in university hospital
- ☆ Treatment of Gynecological diseases like Anoestrus, Repeat breeding, Uterine disease, Dystocia, facility for Caesarean Section.
- ☆ Laboratory testing of blood, urine, faeces, mucous, C.S.F. semen and disease diagnosis through culture sensitivity testing. X-rays, Ultrasound Laparoscopy and Endoscopy.
- ☆ Testing of animals for Brucellosis, Tuberculosis and Johne's disease in animals.
- ☆ Facility for insecticide/pesticide and mycotoxin testing.
- ☆ Brucellosis testing in human beings.
- ☆ Attending to disease outbreaks/epidemics and calamities like fire, flood, electrocution *etc.*
- ☆ Technologies for value addition of milk, meat, egg and fish products for training and also for sale purpose.
- ☆ Mastitis diagnostic kit for farmers and Herbal Teat Dip
- ☆ Milk adulteration detection kit for farmers
- ☆ Publications (Punjabi/Hindi/English)
- ☆ Vaccination and Deworming Calendars
- ☆ Specialized Veterinary Hospital for Dog and Pet animals.
- ☆ Estrus Synchronization protocol
- ☆ Improved germplasm of crossbred cows, buffaloes as semen, embryos, bull calves(HF, Sahiwal and Murrah) and Beetle Goats bucks and Lab animals.
- ☆ Supply of germplasm of poultry like satluj layer, IBL-80 broilers, RIR, Quails, carp fish as well as ornamental fishes
- ☆ Supply of mineral mixture (general, area specific, high yielder and for pigs) and Uromin lick and bypass fat.
- ☆ Formulation of balanced ration for dairy, goat, pig, poultry and other species as per requirement of farmers.
- ☆ Methods of fodder conservation and preparation of concentrate for animals.
- ☆ Analytical laboratory for feed testing.
- ☆ Ideal housing for dairy, poultry, pig, goat, pet animals and ponds for fish farming.
- ☆ Rotatory handy calendars

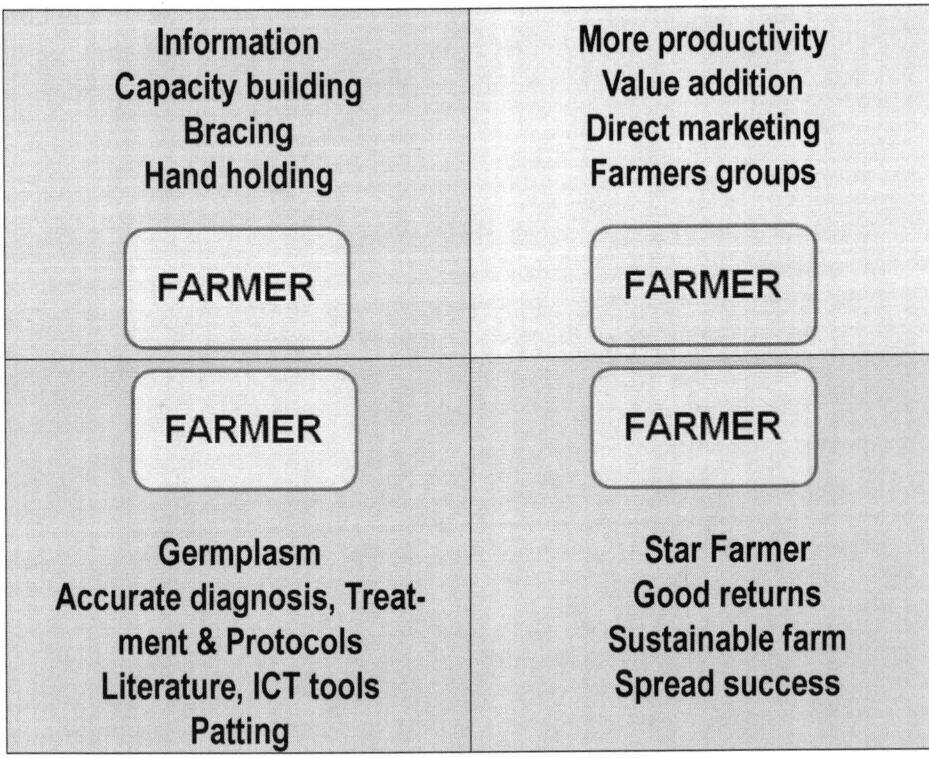

Farmer Friendly Multitier Approach for Better Livestock Productivity

- ✰ Video CD on Repeat Breeding and Anoestrus and their management in Dairy Animals
- ✰ Precision dairy farming App and Dairy Prajanan Mobile App
- ✰ Goat farming and Pig farming mobile Apps.

There is no single perfect technology or procedure which can cater to all the needs of farmers. Technologies must be considered with regards to their total cost for owner. Modern extension tools especially digital ones can save time, man power and money. Extension must be provided on 24x7 basis for the growth development and sustainability. Technology must be user friendly, cheaper, easily updated, accessible and locally available. Technologies must reach to the target people, otherwise they rot lying on shelves or in the books. (Technologies and Fish are always preferred and welcomed fresh)

It always a strong commitment among the faculty and staff to help the farmers coming from all parts of the state and adjoining areas with a trust in the institution. GADVASU, Ludhiana has taken many steps to alleviate sufferings of large, small and pets through 24x7 services for animal owners not from India but abroad also and many animal welfare programmes have been initiated, for making strong bonding between pet and people. These facilities have been developed for the livestock farmers, but they need to contact veterinary varsity.Ultimate beneficiaries are farmers and they must understand the importance of these public institutions and bring their problems to the scientists. The success will only come if their queries are met in time and suitable answers are provided to them and hand holding is done in right direction, this is what GADVASU is up for it, through unique farmer friendly cum multitier approach.

Chapter 14

Synergistic Convergence for Agripreneurship Development and Maximising Farm Profits

Rashmi Singh[1], M.S. Nain[1], J.R. Mishra[1] and Anjani Kumar[2]

[1]Principal Scientists, Division of Agricultural Extension, Indian Agricultural Research Institute, New Delhi

[2]Director, ICAR- ATARI, Patna, Bihar

Agriculture is an important sector of Indian economy as it contributes significantly to the total GDP and provides employment to substantial chunk of the population. Indian agriculture has registered impressive growth over last few decades. There has been 3.7 times increase since 1965 in agricultural productivity through better technology and public investments (*Niti Aayog*, 2017). It provides raw material to all industries and is simultaneously also a market for industrial products. It is also an important earner of foreign exchange. Thus significance of agriculture sector is unmistakably very high.

Present Scenario

There is continuous fragmentation of land due to ownership multiplication and division of joint families into smaller nuclear families, resulting in low per capita availability of land. The land man ratio in our country is becoming very narrow. Due to low land man ratio, more and more farmers and their children are finding themselves out of work. Increasing influence of education has also created a higher need for job amongst the rural masses. After finishing their education, rural youth line up in front of employment exchange for seeking jobs. Although they have an option of starting their own enterprise but this option is usually ignored by them and they join the long queues searching jobs. Another serious problem we have in our hand is rampant unemployment especially in our rural areas. The main cause of unemployment is the rapid growth of population due to which our youth are not finding employment despite their best efforts. As a consequence, they are getting frustrated and migrate to nearby towns and cities in search of petty jobs. The migration presents the cities with challenges and pressure is created on their meager resources. The rural migrants in cities are compelled to live a low quality of life.

Need for Entrepreneurial Agriculture

In view of shrinking land base and rampant unemployment, there is need to commercialise and diversify Indian Agriculture in such a way that one can generate more income per unit of area and time and create agro-based employment opportunities. For this to happen we need to convert agriculture into agri business. With liberalisation, privatisation and globalisation, the whole world has become like a local market where our farmers and their products are required to compete with multinationals in terms of quality and price. Farming as a profession looks dismal and grim. The youth of farm families are not interested in pursuing family profession and are ready to work for menial jobs in cities. Farming has lost its social esteem and is termed as non remunerative.

Agripreneurship-Paradigm Shift Needed

Entrepreneurship Development in Agriculture is an important way out to bring a transformation in our rural areas. Developing Agri-enterprises and making our farmers entrepreneurial is needed urgently to pave the way for evergreen revolution. Unlike crop production, developing entrepreneurship is not an activity but a chain of events and require functional linkages with various support systems. New commodities are perishable in nature and a farmer had to look for marketing channels unlike the established channels as in case of traditional crops of wheat, rice *etc.* In this scenario, a shift in extension approach is also needed.

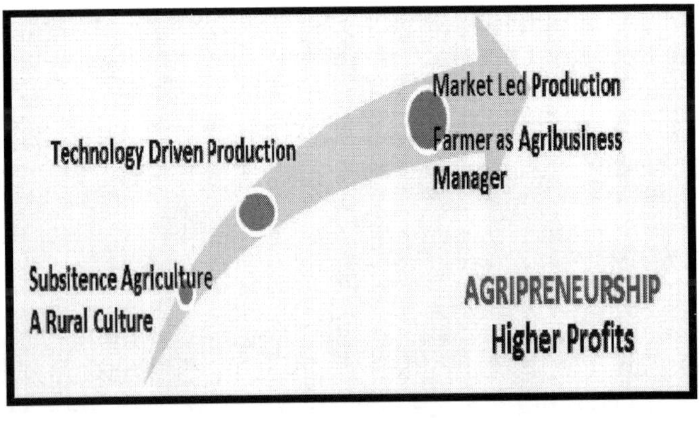

Figure 14.1: Paradigm Shift of Farming towards Agripreneurship.

Agriculture has to move from subsistence culture to Agripreneurship for higher profits. Agri based enterprises are the need of the hour. Agri based micro enterprises can help in poverty reduction, employment generation, empowerment and enterprise development as an end in itself. Regardless of the stage of economic development among the Asia-Pacific countries, small and medium enterprises (SMEs) are generally considered as major sources of employment generation. Promotion of such enterprises, therefore, is generally regarded as part of an employment intensive industrialization strategy.

Focussing Human Domain

It is imperative to focus on human domain in developing entrepreneurship efforts. The neglect of this aspect is proving detrimental to all other efforts directed at setting up agri-enterprises and fuelling entrepreneurial zeal among farmers. Entrepreneurs as the human capital can translate rural economic development. For converting a farm into an enterprise or business, the identity of the person managing it must change from a farmer to that of farm business operator or an agripreneur. There is a great need to make our farmers entrepreneurial. Physical, natural and economic resources are important for facilitating development of any nation but the most important ones are the human resources. Capable, efficient and motivated human resources can bring about tremendous positive changes towards progress. No doubt it is equally true that inefficient people can wither away a whole empire and squander away its precious wealth and abundance. Several empirical studies have shown that the entrepreneurs as the human capital have made a large contribution to economic development than non human capital. Theodore (1970) after comparing the economies of several nations found that it is the quality of human resources which brings about transformation in a society. He contrasted small land area nations like Japan with large land area countries and concluded that it is the entrepreneurial spirit which makes all the difference in the output from the farmers. Developing competent human force is of paramount importance in the present context and entrepreneurial zeal development is the first and foremost step to attain this goal. Increase in population, which is considered a liability can also contribute positively towards overall national development if entrepreneurial qualities are inculcated among the masses.

Developing Agripreneurs: Inculcating Entrepreneurial Competencies

The dimensions of entrepreneurial spirit inculcation include risk taking ability, hope of success, feedback usage, persistence, confidence, knowledge-manageability, achievement motivation, persuasibility and innovativeness. Crop planning all the year around as per the market demand is visualized to be critical in getting higher profits from the same piece of land. Value addition and primary processing are additional options for enhanced farm incomes. Further, well developed inter linkages between agriculture and rural industry will be an effective mechanism of promoting rural transformation. While on one hand farmers get just a few rupees in the exchange of their produce, it is multinationals who are grabbing the major chunk of profits after selling it to the ultimate users. Whether it is potatoes, tomatoes and other perishable commodities, farmers are forced to sell their produce for very less amount while the ultimate consumer pays huge amount when it is processed and packaged after grading. Thus farmer led value addition ventures and marketing are needed for which entrepreneurial competencies must be inculcated among farmers.

Developing motivated, entrepreneurial and efficient farmers requires focus on the human domain, which is the prerequisite to bring about rural transformation The characteristics of entrepreneurs can be developed through systematic motivational training. Many farmers who have been entrepreneurial have attained success and are able to maximize profits from their lands. There are cases of farmers who have also earned Rs 9 lakhs from an acre! The recent recognition of Innovative farmers with Padma Shri Awards by Indian government highlights the potential of farming as high remunerative vocation if pursued innovatively and in a business manner. Entrepreneurship among farmers should be developed through systematic awareness spread programme. Awareness about market potential has to be systematically spread among the rural masses especially the farming community. Then comes the second step of motivation development. This awareness and motivation has to be developed not only in the farmers but also in the field level extension workers. Once they become aware about entrepreneurship opportunities and convinced about its need in agriculture, they will be able to provide some thrust towards commercializing agriculture, as they are the ones who are implementing government sponsored schemes.

Agricultural productivity is believed to be enhanced by the incorporation of strategic entrepreneurship skills. Farmers may be trained to be entrepreneurial to manage agriculture not as a way of life but as an enterprise and trained cadre of youth having capabilities for agri entrepreneurial endeavours. Individual's motivation, aspirations and entrepreneurial competencies trigger agripreneurship development whereas adoption of best practices, dodging of inhibitive factors and facilitative socio economic factors play sequential role in reaching agripreneurial success (Singh *et al.*, 2014).

Agripreneurship Process Model

At IARI, we pursued a study for the last 10 years (2009-2019) on 75 successful agripreneurs and delineated key Agripreneurship Process Catalysts as individual motivations, aspirations, entrepreneurial competencies, facilitative socio-economic factors and also their ability to effectively manage the inhibitive factors. Facilitative factors for their success were found to be enough available opportunities, market demand, proximity to markets, technological updation/expert advices, grouping for raw materials and utilizing low cost modified machineries. The best practices which achiever farmers were engaged in are: Diversification, Innovativeness, Market oriented crops, effective marketing linkages, capitalizing prior experience and quality concern.

On the basis of analysis of the successful cases and the action interventions undertaken a framework for agri-entrepreneurship and farmers' innovation dynamics has been conceptualized. The agri-entreprise development for maximizing farm profitability was found to be interplay of entrepreneurial competencies, entrepreneurial climate, and farmers' innovations (Figure 14.2). It was found that the competencies like opportunity recognition, drive for excellence, quality concern, moderate risk taking behaviour, innovativeness and business orientation in presence of suitable climate like networking, infrastructure, government priority and financial backstopping lead to experimentation not only for technological innovation, but also new ways of managing livelihood in general (networking, communication, institution building, information management, marketing, planning, accessing resources, *etc.*). The innovations which were economically viable and found sustainable source of funding were able to translate into entrepreneurial ventures having higher income and profits. Social networking of farm innovators has proved to be potential to construct knowledge. On the other hand to maximize the income, the farms required certain distinguishing capacities like foreseeing institutional requirements and linkages, comparative financial impact and success analysis ability in addition to analyze projected demand and required changes in socio cultural and infrastructural domain. The results are in conformity with Singh *et al.* (2014, 2016).Whereas, it was inferred that individual motivations and aspirations trigger entrepreneurship and the competencies along with best practices (innovations) and convergence of synergistic linkages play sequential role for enterprise success.

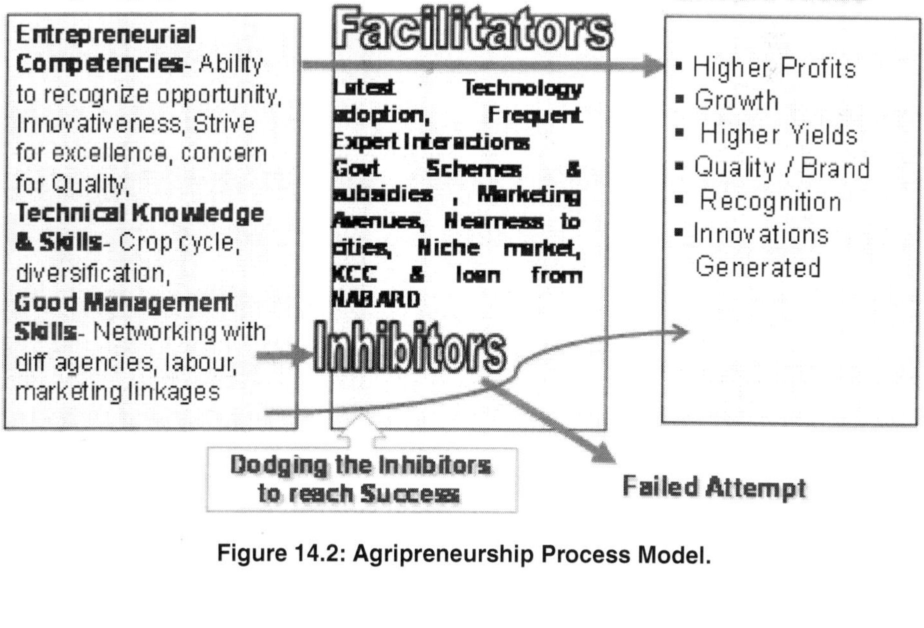

Figure 14.2: Agripreneurship Process Model.

Convergence of Synergistic Networking for Farmer based Agri Enterprise Development: Experiences of Action Research

Research studies undertaken by IARI during last ten years have revealed that it is possible to build entrepreneurial competencies of our farm women through appropriate training interventions, mentoring and linking them with other agencies. As per the action interventions undertaken training course were organized, follow up and mentoring of the trained youth was done and linkages initiated with supportive institutions and agencies. Farm Women who have been trained earlier and have launched their pickle making, *dalia* making, wheat flour mill and grinding of spices enterprise named "*Sunhara Lalpur*" and "*Neelraj*". Handholding and mentoring activities were undertaken to establish their enterprises on a strong footing. Linkages with other agencies were established for finance requirements, marketing and certification purposes.

Under collaborative training course by IARI and RUDSET Institute, twenty eight women entrepreneurs from three project villages of Hapur were trained in Baby corn cultivation and its processing, Post-Harvest Processing of soyabean, bajra puffs, crackers *etc.*, Value Addition of Vegetables/Fruits and Floriculture and dry flower technology for enterprise uptake. Convergence of Synergistic efforts of all stakeholders for farmer based agri enterprise development was taken up. Farm women entrepreneurs of Hapur villages were linked with all stakeholders in the process- IARI for mobilizing and mentoring, RUDESETI for training, Agricultural Systems India Ltd (ASI) and Bharati Walmart for marketing, NGO *Laxmi Jan Kalyan Sewa Sansthan* for grassroots mobilization and logistics and NABARD for financial support. The initiative resulted in three marketing outlets under the name of Sunhara Fresh in Hapur, Two outlets for Agri-inputs, Fair price shop as a non farm venture launched by the rural women of project village Shyampur.

Institutional Convergence: Way Ahead for Developing Agripreneurship

What do we mean by convergence? Convergence literally refers to the alignment of issues, interests and therefore solutions across all institutions engaged in development of Agripreneurship. Thus convergence is not just collectivization, it is not referring to cooperation only and it is definitely more than collaboration. In fact, rather than just being episodic and event-driven, convergence actually implies constant shared commitment with recognized "wins" for all partners involved. The emergence of multi-stakeholder alliances between profit oriented companies, social enterprises, non-governmental organizations (NGOs) and government institutions collaborating in totality is an important solution to many of our problems plaguing agriculture sector. This structure calls for new ways of thinking crossing traditional organizational, functional and sectoral boundaries of various institutions. Also the most important thing being able to develop mechanisms for increased dialogues, shared learning and joint monitoring of outcomes as well as processes.

REFERENCES

High Level Meeting Communiqué, 16 April 2014: First High-Level Meeting of the Global Partnership for Effective Development Co-operation: Building Towards an Inclusive Post-2015 Development Agenda Mexico

http://en.wikipedia.org/wiki/Grameen_Danone

Singh Rashmi, *et al.* (2012): Entrepreneurship Development in Agriculture, Biotech Publishers. Project Reports, Enhancing Entrepreneurship among Rural Youth, IARI- P.I. Rashmi Singh.

Singh Rashmi, Nain, M.S. and Mishra, J.R. 2014. Institutional Convergence of Synergistic Strengths for Developing Women Agripreneurs. *Indian Journal of Extension Education*, 50(3 and 4):1-7

Yunus, Muhammad (2010). Building Social Business. New York: Public Affairs. p. 226. ISBN 978-1-58648-824-6.

Chapter 15

Extension at the Crossroads!?

Sujeet K. Jha

Principal Scientist (Agricultural Extension),
ICAR-National Dairy Research Institute, Karnala – 132 001, Haryana

Introduction

On account of inculcating the sense of belongingness to the discipline of *extension* since more than three decades, it gives me an immense pleasure to be a part of this august gathering, on the occasion of **"ISEE National Seminar-2019"**, being organized by the Indian Society of Extension Education (ISEE). In fact, I really feel privileged that, the organizers of this event have provided me with this opportunity of having interactions with you through this presentation of mine, in the form of *"Lead Paper"*, while sensitizing the fellow *extensionists* vis-à-vis deliberations on the issue of '*Extension at the Crossroads!?*' Will be done. With particular reference to '*Extension* Discipline'. Hopefully, this should help in the process of generation of ideas relevant to this issue via '*Brainstorming*' exercise(s) among the delegates, who have come here from different corners of India to participate in this great event. Accordingly, various issues of concern have been discussed as below:

Extensionists* being treated as a sort of *punching bag* for poor or low rate of Adoption of the recommended Technologies/Practices among the intended *End-Users

In India, the extension personnel, in general, get targeted and made the '*scape-goat*' on account of poor or low rate of adoption of the recommended technologies/ practices ! Isn't it a fact that...given the poor infrastructure, coupled with limited resources available at the disposal of the extension personnel of a country like ours... they (the extension personnel) involve themselves, whole-heartedly, towards the cause of *Transfer of Technology* (*ToT*)... wherein, not always, they get the due success; albeit, the reasons behind the poor rate of success could be manifold!? Moreover, the interesting thing is, that...if the *Technology* concerned happens to get adopted at the field-level, among the intended end-users...the whole *credit* goes to the Scientist(s), who developed the said *Technology* ; whereas, if due to some reasons, the same *Technology* does not get accepted and/or adopted by the targeted audience/ beneficiaries... the entire blame is put on the extension personnel involved in the said process, thereby treating them as a sort of *punching bag*, on account of the failure of the *Technology* concerned! Why it is like that !? Why nobody questions those people who developed the said *Technology*!? Why no one tries to understand the fact that... may be, the *Technology*, in question, might have failed in terms of the attributes like: feasibility, complexity, compatibility, relative advantage, *etc*....as perceived from the perspective of the intended end-users!? So, in my opinion, this kind of *mind-set*, as existing among most of our fellow colleagues belonging to *Sister Disciplines* (under the domain of Agriculture and allied sectors) needs to be changed!!

Poor level of 'Perceived Credibility' of Extensionists among our Fellow Colleagues belonging to other Disciplines within the ICAR/SAUs system

In fact, at this juncture, it would be worthwhile to note that...whenever we, the Scientists, Faculty, and Research Scholars/Students belonging to *Extension* discipline happen to interact and discuss with the people belonging to other professions and/ or occupations, especially those, who are from outside the ICAR/SAUs System, they get very much impressed with the nature of our job; and also, they genuinely give the feedback and/or feel that...we people (the extension personnel) are engaged in some sort of altruistic endeavour vis-à-vis enhancement in *'Quality of Life'* of the intended clientele, thereby leading towards overall welfare of the poor people of our country ! Rather, many of such people do opine, categorically, that...the *extensionists* carry out the kind of tasks *a la* professionals having degrees like: Master in Business Administration (MBA) !! However, unfortunately, the same *extension personnel* are mostly perceived as *"good for nothing"* by our fellow colleagues belonging to other disciplines within the ICAR/SAUs system !!! The reasons behind this behaviour on the part of our fellow professionals happen to be beyond comprehension !? Nevertheless, this issue needs serious and analytical thinking.

Encroachment by the *Outsiders* on the Job-opportunities offered by our *'Discipline'*

Further, to make the matter worse...due to some unknown reasons, many people belonging to other disciplines have had been allowed to enter our *'Discipline'*, on many occasions...and, as a result of it, present-day Scientists, Faculty, and Research Scholars/Students belonging to *Extension* discipline do feel that their *'discipline'* as well as *'profession'* seem to have *'diluted'*, to some extent, definitely, if not *in toto* ! At the outset, this statement may seem to be acrimonious in nature, but this happens to be very much true...as this "policy" has proved to be "Extension's *Achilles' heel*" since long...as many people from other *disciplines* are encroaching on the job-opportunities offered by our *'Discipline'*, *viz.* "Director of Extension" in various SAUs; "Joint Director (Extension)" at the National Institutes of ICAR; Directors of the I.C.A.R. Institutes like: Agricultural Technology Application Research Institute (ATARI), which was known as Zonal Project Directorate (ZPD);

Programme Co-ordinators of *Krishi Vigyan Kendra* (K.V.K)., *etc.* Further, *to add insult to injury*, even the people from other disciplines are being encouraged by the ASRB to act as an "*Extension Expert(s)*" during the interviews conducted by ASRB to select "*Extension Scientists*"!

'Identity-crisis' and/or Confusion among Extension Scientists vis-à-vis '*Role-Expectations*'

Sometimes, there seems to be some sort of 'identity-crisis' and/or confusion among extension professionals vis-à-vis their '*role-expectations*'. To make this point more clear, let us consider the concepts of '*Centre-Extension*' as well as '*Field-Extension*' ... wherein, in the case of former, the *extensionists* are supposed to work and contribute towards the cause of '*Extension*' as a '*Discipline*'; and, in the case of latter, the *extensionists* are supposed to function as the field-level extension-workers, in order to contribute towards the cause of 'Extension' as a '*Service*'. However, the confusion arises, only when the people associated with the former, especially at organizations, like – ICAR Institutes/Deemed (to be) Universities as well as SAUs are expected to do the job of the latter ... and, to make the matter worse, the situation becomes more confusing due to the fact that: the 'professionals' associated with '*Centre-Extension*' already know that, by contributing for the latter, they would not do any good for themselves, career-wise, as per the established norms of further promotions; and, therefore, they show the feeling of 'reluctance', on their part. Hence, it is essential to find out a 'way out' to mitigate this issue/problem/confusion...especially when, most of the time, the *services* of the "*Extension Scientists*" are required mainly for organizing some kind of events at the field-level !?

'*Extension Scientists*' not being preferred, generally, as '*Partner(s)*' for "Inter/ Multi Disciplinary Research Projects" :

It is widely believed that...every research project/programme meant for developing the technology(ies) to be transferred to the field-level must involve *extensionists*, in order to obtain their views about the socio-cultural perspectives involved in the transfer of such technologies among the intended clientele. Accordingly, it would be imperative to test the feasibility of any new technology/ technique (developed by the researchers) at the level of research-institution itself ... and, this will have to be done by the *extensionists* alone ... before transferring such technology/technique to the field-level.

However, unfortunately, despite the cause of "*inter-disciplinary* and/or *multi-disciplinary* researches" being advocated, time and again, very few of the *extensionists* are enlisted as the "*partner*" in such endeavours ... the reason, behind it, being beyond one's comprehension ! May be, the researchers do not want to be "distributed" in their mission of developing varieties-after-varieties, on regular basis, without ever considering the practical feasibility of '*adoption*' of such 'technologies' at the actual field-level !!

Ignoring the '*Extensionists*' for "Research and Management Positions (RMP)" at the apex level on account of their poor 'Score Card':

Of late, there has been a lot of hue and cry on account of the system of '*Score*

Card' developed recently by the ICAR/ASRB ... wherein, the field-related and/or 'Extension' oriented activities have been almost overlooked/neglected, thereby resulting in resentment among the *'Extensionists'*; since majority of them find themselves unable to compete with the specialists belonging to other disciplines, while applying for the RMP posts.

In fact, this matter had been even taken up and discussed by the then Hon'ble Union Minister of Agriculture and Farmers' Welfare, Govt. of India as well as by the then Hon'ble Minister of State (Agriculture and FPI), Govt. of India ... in the august presence of the then Chairman, ASRB as well as Secretary (DARE) and Director General, ICAR ... during an event organized at ICAR-National Dairy Research Institute (NDRI), Karnal (Haryana), in JAN.2015. In that meeting/event, both the then Hon'ble Ministers raised this issue, and argued on behalf of the *'Extensionists'* that...all the field-related activities being carried out by the extension personnel must be given some sort of weightage (in terms of 'Scores'), appropriately, so that the *'Extensionists'* should not have any kind of grudge on account of obtaining lesser 'Scores' as compared to their fellow colleagues belonging to other *'Disciplines'*! However, unfortunately, nothing concrete has been done, so far, in this regard!!

Multitude of 'Societies' in the name of '*Extension*'

As of now, there happen to be a large number of *'Extension Societies'* existing across our country, *viz. Indian Society of Extension Education, Society of Extension Education, Maharashtra Society of Extension Education, etc.* ... thereby indicating about the absence of any 'Single' Society of 'Extension' discipline looking after the affairs of 'Extension Discipline' vis-à-vis *'Formulation of Policy and Guidelines'* which could be implemented, uniformly, throughout the country ! Therefore, it is high time that ... a 'Single' Society of this discipline should be there, at the national level, in order to provide a sort of 'umbrella' for the *'Extension Professionals'*; and, which, in turn, would help all of us to get united and stronger. In fact, one feels that ... almost many of the afore-discussed issues would get resolved, instantaneously, if there would be only one 'Society of Extension Discipline' at the national level !?

Lack of Good Quality of Journals in the Discipline of Extension vis-à-vis NAAS Rating

It is very sad to note that...none of the *Journals* of *Extension discipline* (including the first and foremost *Journal* of *Extension discipline,* named as : *Indian Journal of Extension Education* (IJEE), being published since more than five decades) have good ratings by the NAAS, as of now...as the ratings of these *Journals,* as per NAAS, happen to be even below 6.0, which happens to be a common/minimum yardstick for the *Journals* of other *disciplines !* Hence, this matter needs to be considered on a priority basis...in order to safeguard the interests of Scientists as well as Researchers/Scholars/Students of *Extension discipline !!*

Need to inculcate the sense and feeling of 'Loyalty' towards our own 'Discipline'

And, last but not the least, we must inculcate the feeling of loyalty and respect to our discipline/profession...since unless and until we *"Extensionists"* would love our discipline/profession from the core of our heart, and fight together for furthering its cause, the "outsiders" (*the people belonging to other Disciplines*) will continue to intrude into the domain of our discipline/profession ! However, for that to happen, we need to communicate to the people at the *helm of affairs*, that... there is a need to motivate and encourage those having 'Specialised Education in *Extension* Discipline' at the time of selection/recruitment of suitable candidates for the posts of Directors of the I.C.A.R. Institutes like: Agricultural Technology Application Research Institute (ATARI), which was known earlier as Zonal Project Directorate (ZPD). Similarly, the posts like: Directors of Extension (in the SAUs) as well as Programme Coordinators (in the KVKs) across the whole 'Agricultural System' (including ICAR/SAUs, *etc.*) should also be filled-in by those persons only, who have done 'specialisation' in the discipline of '*Extension*'!!

Conclusion

Finally, on the basis of afore-discussed issues, it may be argued that... although *much water has flowed under the bridge* since the introduction of "*Extension*" as a "*Discipline/Subject*" in India (at Bihar Agriculture College, Sabour, Bhagalpur, Bihar, in the year 1955), now, even while almost six and a half decades have passed since the inception of this *discipline*, in our country ; it is high time that : in the changing global scenario, the '(*Indian*) *Extension*', in terms of both '(*Educational*) *Discipline*' as well as 'Rural Service(s)' alike, evolves itself into such a way that the perceived credibility of "*Extension*", among the public, at large, could be enhanced, significantly, thereby, minimizing the probability of "*Extension Professionals*" being blamed for poor or low rate of adoption of the recommended technologies. In fact, till the time 'food' cannot be 'downloaded' for the eating purposes, it may be presumed and/or argued that...the '*Discipline*' of "*Extension*", *per se*, holds a lot for the future ahead, since the burgeoning population in a country like ours have to be fed, sustainably; and, for that to happen, the '*technologies*' being developed under the domain of *Agriculture and allied sectors* need to get adopted at the field-level, among the intended end-users...albeit, without involving the expertise of the *Extension Professionals* in the entire process of ToT, the objectives of developing all such '*technologies*' might not get fulfilled !? Besides, the *extensionists* must also facilitate in encouraging and motivating their intended clientele/end-users/target-audience to participate actively and effectively in the process of their own welfare and development. And, that means...the *discipline* of *extension* has a prominent role to play vis-à-vis upliftment of rural poor as well as enhancement in their '*Quality of Life*', in the long run...albeit, their efforts in making this endeavour successful need to be appreciated by one and all, across the society, without any bias and prejudice. In this context, for furthering the cause of '*Discipline*' of "*Extension*", even the services of the *alumni*, belonging to the discipline of "*Extension*" *and* holding good positions in different organizations may be taken/sought...as, they might contribute as the '*Brand Ambassadors*' for our '*Discipline*', by virtue of their own '*credentials*' as well as the '*goodwill*' generated among the peer group, in particular, and the society, in general !?

Chapter 16

Nurturing Entrepreneurial Ecosystem in Agricultural Universities

Surya Rathore

Principal Scientist (Extension Information Systems),
ICAR – National Academy of Agricultural Research Management,
Hyderabad – 500 030, Telangana

"A business ecosystem is just like the natural ecosystem; first, needs to be understood, then, needs to be well planned, and also needs to be thoughtfully renewed as well."

— Pearl Zhu

An Entrepreneurial ecosystem is made up of a set of interdependent components and factors which work in harmony so as to enable fruitful entrepreneurship within a specific boundary. The elements of an entrepreneurial ecosystem could be individuals, groups (a group of students, Farmer Producer Organisations) and institutions such as colleges, Technology Incubation Centres which form a community by interacting with one another. Apart from this, some environmental determinants which influence the ecosystem are laws and policies that govern the individuals, groups or institution. Very recently, there was a news that IIT, Madras and Phoenix Medical Systems have launched a standing wheelchair. This is what happens when innovations and opportunity meet. Not only this, IITs have a culture of innovations and technology commercialization. Why can't we find this kind of scenario in our State Agricultural Universities? To create this kind of culture, we need to have an ecosystem which is conducive for entrepreneurship or in other words, the right kind of environment for sprouting innovation culture which would lead to entrepreneurship. The environment, the people and the institutions all make the ecosystem which is fruitful for entrepreneurship.

Why do we Want our Agri-graduates to be Entrepreneurs?

As there is a very thick line between number of Agri-graduates passing out every year and job opportunities available for them in the real time market which forms an utmost requirement for creation of jobs to the rising number on a large scale which is technically a mammoth task for any country's economy. According to an estimate (IAMR, 2001), 43 per cent of the graduates and 23 per cent of postgraduates (M. Sc.) face the challenges of employment in our country. The only and most reliable option remains for many of them is to earn their livelihood by entering into the field of entrepreneurship. India has a huge potential for entrepreneurial activity given its sheer demography. Entrepreneurship education here can be used as a tool to channel the untapped human resources and produce a large number of job creators rather than job seekers. Colleges need to encourage students to become job providers by making them learn to challenge the status quo, taking risks. Even, Government of India is providing necessary support for levelling entrepreneurial education among different strand of people and supporting them technically and financially.

Present Context

Taking an example of 60 agripreneurs graduating from five State Agricultural Universities of the country, the entrepreneurial ecosystem (World Economic Forum 2012) is explained thoroughly and based on the findings, a model is suggested for our State Agricultural Universities.

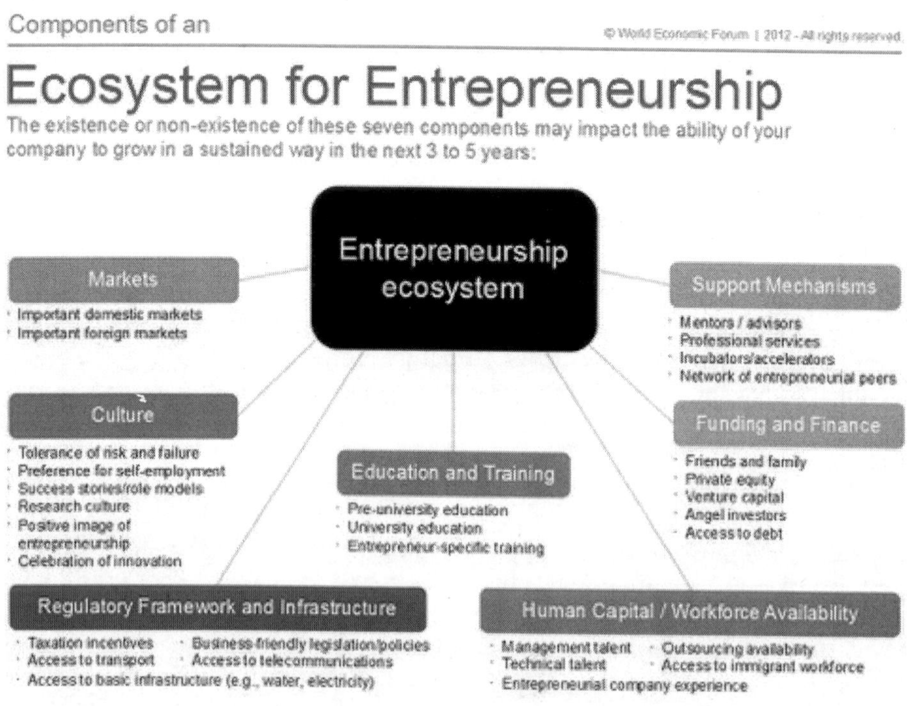

***Source*: World Economic Forum (2012).**

Components of Entrepreneurial Ecosystem

According to World Economic Forum (2012), Entrepreneurial Ecosystem includes the following components:

Markets

A market is any place where makers, distributors or retailers sell and consumers buy. Businesses that operate in markets are usually in competition with other companies. Three basic types of market structures are (1) Perfect competition: many buyers and sellers, none being able to influence prices. (2) Oligopoly: several large sellers who have some control over the prices. (3) Monopoly: single seller with considerable control over supply and prices. The agripreneurs under study were the ones who had experienced two basic types of situations that is oligopoly and monopoly. More than half had to depend on market price for selling their own products whereas others used their own pricing strategy. It is quite obvious that monopoly can only exist where there is innovation. Present study 80 per cent of the entrepreneurs under study boasted that there was some innovativeness in their business. For example making of '*Panchamratam*', an organic manure of cow's urine and four types of leaves, Bio compost screening assembly, Designing and manufacturing of low cost machines for packaging of pickles, balancing of colour value and pungent value in chilli powder are some of the innovations practiced by entrepreneurs under study.

Culture

The question arises here is do we have a culture of entrepreneurship in our agricultural university system?

The word culture here refers to having tolerance to risk and failure, preference to self – employment, having success stories and role models, research culture, positive image for entrepreneurship and innovations welcoming atmosphere. For example ICAR has introduced courses like entrepreneurship development in the 5th Deans' committee recommendations which would up to some extend inculcate the entrepreneurial seeds among the Agri graduates. The student ready programme at bachelor's degree level too is a welcome step in this direction. We at ICAR – National Academy of Agricultural Research Management, Hyderabad have documented more than 30 success stories of agrigraduates who have turned into entrepreneurs in book form and also developed 5 videos covering five State Agricultural Universities of the country. These videos are put on YouTube and can be shared with budding agrigraduates. To welcome innovations, Agribusiness Incubators is a vehicle which promotes agricultural technologies developed by R&D centres of excellence, universities and other institutions, separately and jointly. Some examples are: NAARM – Agri Business Incubator (ABI), Hyderabad; Pusa Krishi Incubtor at Indian Agricultural Research Institute, New Delhi, Agri Business Incubator – International Centre for Research in Semi-Arid Tropics (ABI – ICRISAT), Pathancheru. All these support Agri-startups in scaling up their business and commercialize their products. Apart from this, ABIs are also present in some of the Agricultural Universities of the country.

Regulatory Framework and Infrastructure

This include taxation incentives, access to transport, business friendly legislations and policies, access to transport, access to telecommunications and access to basic infrastructure. Any start-up that has been incorporated after 1 April 2016 can get a 100 per cent tax rebate on its profits for a total period of 3 years within a block of 7 years. However, if the company's annual turnover exceeds Rs 25 crore, then the tax rebate is not valid. This tax exemption has been put in place to help businesses meet their capital requirements while setting up.

Education and Training

This pertains to pre-university education, university education and entrepreneur specific training. In India, agriculture as a subject is not taught in all schools barring a few. In university education, we have 70 plus agricultural universities along with a good number of private players. Coming to training, there is a provision of Agri Clinic/Agri Business Centres providing training to agricultural graduates by means of Nodal Training Institutes spread throughout the length and breadth of the country. Also, Agricultural Skill Council of India (ASCI) is in place. To accomplish its task of skill upgradation and employment generation, ASCI has to collaborate with all key stakeholders including Agricultural and Veterinary universities and related research institutions. It has Memorandum of Understanding (MoUs) with some of the Universities so as to align the short-term skill training programmes of universities with the Qualification Packs developed by ASCI which no doubt falls in line with National Skill Qualification Framework of Government of India. ASCI develop the skills and then connect the trainees to various companies for employment.

Support Mechanisms

Presence of mentors, advisors, professional services, incubators, accelerators and network of entrepreneurial peers. The government has launched an Agriculture Entrepreneurship Promotion Scheme (AEPS) to assist the agriculture entrepreneurs across the rural areas in the country. The scheme was initiated during 2018. In an attempt to promote innovation and entrepreneurship in agriculture, the ICAR - NAARM Technology Business Incubator (TBI), has launched a new AGRI-UDAAN programme that will help for the innovative start-ups in Agri value chain for effective improvement in agriculture. In order to promote a culture of innovation and entrepreneurship, Government of India launched Atal Innovation Mission (AIM) with an objective to serve as a platform for promoting world class start-up businesses, innovation hubs and technology driven self-employment activities. Under Self Employment and Talent Utilisation (SETU) component, entrepreneurship is promoted through self-employment and talent utilization. Under the component "Innovation promotion" a platform for generation of innovative ideas is provided. Agricultural and Processed Food Products Export Development Authority (APEDA) is an apex body under the Ministry of Commerce and Industry, Government of India, responsible for the export promotion of agricultural products. Majority of the agripreneurs under study said they had friendly relations with their peers in the same business and around a quarter of them were acquiring knowledge from peers. So, it is deemed necessary that entrepreneurs should have awareness of the support mechanism available for them.

Funding and Finance

Getting finance for establishment of business is a very vital component to give spark to one's entrepreneurial dreams. One can get financial support from friends and family, venture equity, venture capital, angel investors *etc.* In the study, maximum portion of respondents (43 per cent) had used multiple sources of finance from family and friends as well as private financers followed by self-savings (27 per cent), family finance and bank loans. Agri Clinic/Agri Business Centres (AC/ABCs) as well as incubation centres all introduce their trainees/incubates with various banks for the purpose of finance.

Human Capital/Workforce Availability

Management talent, technical talent, outsourcing availability, access to workforce and entrepreneurial company experience; all are included in this. The success stories of the study revealed that technical talent and management talent are both equally important which can practically be gained through working somewhere as a professional and gain experience of the respective domain. As the saying goes, **'practice makes a man perfect'**. With all the knowledge, more than 40 per cent had international business. All the successful entrepreneurs under study knew both Hindi and English languages and majority of the entrepreneurs linked language knowledge with better customer retention and collaboration with co-workers.

Now, the question arises as to how to develop a startup ecosystem in our State Agricultural Universities? For this, first of all agriculture subject should be made compulsory at school level and students should have the option to have agriculture as their optional subject, though some of the states including Maharashtra and Rajasthan have; but not in all schools. This would inculcate a kind of affiliation with the discipline. Agricultural Education division of the ICAR should establish Vocational Counselling Cells in Agricultural Universities and curriculum comprising Agricultural Marketing, Trade and Prices should be broadened a bit where people starting new businesses and existing business visionaries struggle more often. The subject of Entrepreneurship Development and Business Communication which was recently added in curriculum should be focused more by having interactions between students and Agri-graduates turned entrepreneurs. Home Science which is also called as Community Sciences should also encourage some courses on Agricultural Marketing. New entrepreneurial ventures should be identified in various Agricultural Sciences and there should be a Vocational Counselling Cell within each college limits for entrepreneurial development.The Placement cells should handle the responsibility of exposing budding Agri-graduates to entrepreneurship core and should be indulged in entrepreneurial ventures supporting activities and handholding for a few years. Information Technology aspect to be strengthened in the curriculum and Agri-graduates should be taught the importance of Information Technology usage in Agriculture such as business marketing by means of Digital Marketing. Innovation knowledge should be promoted among Agri-graduates and the phenomenon of awarding candidates with suitable innovative ideas to be endorsed in the SAUs. There should be convergence of Agri-Clinic/Agribusiness scheme of MANAGE, Hyderabad with SAUs and establishment of linkages with EDI (Entrepreneurship Development Institute of India), Ahmedabad and Agri-Business Incubators of ICAR.

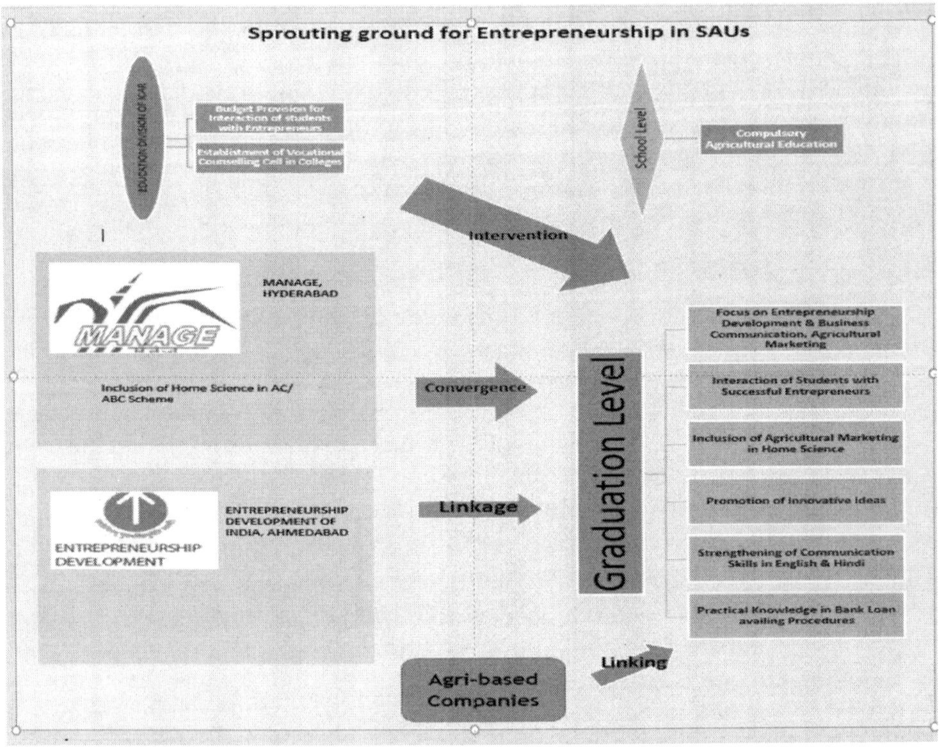

Suggested Model for Developing Startup Culture in SAUs.

The Success Stories of Agri-graduates should be shared with budding Agri-graduates by means of providing them with literature of success stories, video films and Face-to-Face interactions with successful entrepreneurs. These stories should also be shared with industry giants which can form a linkage between them and Upcoming Agri-graduates and may result in few fruitful ventures in the near future. Languages should be given a special attention and Hindi and English languages must be made compulsory throughout the country's educational limits in order to boost business marketing capabilities of upcoming entrepreneurs and should be included in Fifth Deans Committee Report curriculum. Importance should be given to banking knowledge and filling up the financial need of the upcoming graduates by making them aware of the financial policies of the country and processes involved in retrieving loan facilities and other important requirements. It was revealed through documentation of success stories of Agri-graduates turned entrepreneurs that there was some sort of institutional backup behind their success. Besides the entrepreneurial spirit of an individual or group of individuals, the enabling environment supporting these initiatives is of utmost importance. Such an environment can only exist in the presence of certain policies and institutional support which fosters startup culture. This type of startup ecosystem if attempted in agricultural sciences would act as a building block for the budding Agri-graduates to enter into the path of entrepreneurship and fulfil the vision of our Prime Minister; Mr. Narendra Modi of having "Job Creators" and not "Job Seekers".

REFERENCES

Lans, T., Verstegen, J. and Mulder, M. 2011.Analysing, pursuing and networking: a validated three-factor framework for entrepreneurial competence from a small business perspective. *Int Small Bus J.*, 29(6): 695-713.

Movahedi, R., Saadi, H. and Yaghoubi Farani, A. 2011. Creating linkages between the labor market and agricultural higher education in Iran Strategies and mechanisms for partnership. *Industry and Higher Education*, 25(4): 307-317.

Sabouhi, F. 2000. Surveying and analysis of unemployment reasons between graduates in Higher Education. Proceeding of need assessment projects on skillful human resources, institute of research and planning of higher education publishers, Tehran, Iran.

Shahbazi, E and Ali-beigi, A. **2007.** The Required Competencies of Agricultural Graduates for Entering Job Market. *Agricultural Extension and Education Science Journal*, 2(1): 14-24.

Verhees, FJHM, Kuipers, A. and Klopcic, M. 2011. Entrepreneurial proclivity and farm performance: the cases of Dutch and Slovenian farmers. *Int J EntrepInnov*, 12(3):169-77.

Zamani, G.H. 2001. Knowledge and technology transfer, a case study: linkage between college and agriculture of Shiraz University and extension service in Fars province. *Journal of Science and Technology of Agriculture and Natural Resources*, 4(4): 57-65.

www.startupindia.gov.in

Chapter 17

Climate Change Workforce Development for Extension

Kalyan Ghadei

Professor, Department of Extension Education, Institute of Agricultural Sciences, BHU, Varanasi – 221 005, Uttar Pradesh

Introduction

The intergovernmental panel on climate change (IPCC) which takes account of on historical, psychological, social, physical and economic and political perspectives of climate change in its report, released in November 2014 said that the Working Group I contribution to the IPCC's Fifth Assessment Report (AR5) considers new evidence of climate change based on many independent scientific analyses from observations of the climate system, paleoclimate archives, theoretical studies of climate processes and simulations using climate models. Climate change is not controversial as naysayers and some media would convey. After decades of careful observation, collection of data, and tracking of changes in the climate system, there exists a solid, scientific consensus that human-caused climate change is a reality (Michael, 2012). In their report, the NCSE 2015 says that there is substantial scientific agreement about the occurrence, causes, and consequences of climate change. Yet due to the inherent complexity of the topic and the social controversies surrounding it, confusion and doubt often persist. Here, we need to develop the workforce, education, and training for climate-smart Agriculture. The following are some interventions.

Creation of Awareness

From the discussion made above and review of research it clear that a thorough understanding on climate change is need of the time. It calls for massive level action plan for awareness of different stake holders in Agriculture such as the students, farmers, agri-prenuers, scientists, extension agents and the general public as a whole. A US study reports that five factors that should account for changes in levels of concern: 1) extreme weather events, 2) public access to accurate scientific information, 3) media coverage, 4) elite cues, and 5) movement/counter movement advocacy. A time-series analysis indicates that elite cues and structural economic factors have the largest effect on the level of public concern about climate change. While media coverage exerts an important influence, this coverage is itself largely a function of elite cues and economic factors. Weather extremes have no effect on aggregate public opinion. Promulgation of scientific information to the public on climate change has a minimal effect. The implication would seem to be that information-based science advocacy has had only a minor effect on public concern, while political mobilization by elites and advocacy groups is critical in influencing climate change concern (Brulle, Carmichael, and Jenkins, 2012). Media coverage of global warming had an immediate but short-term influence on public concern. With such transitory high levels of media coverage we suggest that for more effective communication of climate change, strategies aimed at maintaining mass-media coverage of global warming are required (Yuki and Midori, 2008).

Incorporating Climate Change Issues in School Education

Since the climate is changing we have to bring the issue to the students and the educational system. A well-planned curriculum from school to university education is a need to educate different stakeholders about the impact, adaptation, and mitigation and to bring a profitable climate resilience crop production. The NCSE 2015 urges "It is important for the science of climate change to be taught, both in formal and informal educational environments, in order for future citizens to be able to make scientifically informed decisions about the consequences of climate change. But educators face challenges in helping their students attain climate literacy". The reality of global climate change lends increasing urgency to the need for effective education on earth system sciences, as well as on the human and behavioral dimensions of climate change, from broad societal action to smart energy choices at the household level. It is noteworthy to view of the US government for climate education that explains "the reality of global climate change lends increasing urgency to the need for effective education on earth system sciences, as well as on the human and behavioral dimensions of climate change, from broad societal action to smart energy choices at the household level (NRCNAS,2011).There are other numerous voices from the organization such as UGCRP, ACS, and UNESCO that support both formal and informal climate change education and climate change literacy(NCSE 2015). The IPCC report (2019) finds that strongly reducing greenhouse gas emissions, protecting and restoring ecosystems, and carefully managing the use of natural resources would make it possible to preserve the ocean and cryosphere as a source of opportunities that support adaptation to future changes, limit risks to livelihoods and offer multiple additional societal benefits. IPCC report on October 2019 also describes fellowships and scholarship for motivating students on climate change education.

Workforce Development

Climate change is multidimensional. It needs a workforce of different categories even in Agriculture. We need experts of different types, including climate specialists in every subject in agriculture to overcome climate change. Understanding and responding to climate change, and make suggestions on how we can use the strengths of the human mind and social dynamics to communicate climate change in its temporal context. Then we propose an integration of these insights for the problem of climate change and discuss mismatches between the human mind, surrounding social dynamics, and climate change (Pahl, Sheppard, Boomsma, and Groves, 2014). The demand for clean energy workforce development programs is becoming increasingly important at both the state and local levels. Appropriately trained workers will be needed to transition to a cleaner economy that addresses climate change and energy challenges while also creating new jobs. Some of these new jobs will require little additional training while others will require training or possibly advanced degrees (EPA held a webcast on this topic in January 2010). Building human capacity for improved assessments requires expansion of skills within the existing public and private sectors and developing a much larger workforce that excels at critical and interdisciplinary thinking. The US global change research program suggests useful capacities include the ability to facilitate and communicate research and practice, manage collaborative processes to allow for imaginative analysis and solutions, develop sustainable technologies to reduce climate risks, and build tools for decision-making in an internationally interdependent world. A deeper understanding of the processes and impacts of climate change, disaster risk reduction, energy policy impacts, ecosystem services and biodiversity, poverty reduction, food security, and sustainable consumption requires new approaches to training and curriculum, as well as research to evaluate the effectiveness of different approaches to research and teaching (NCA 2014). The Paryavaran Mitra (Friends of the Environment) in India is an action-based education program for children with the spirit of increasing the Hand Print focusing on Sustainability and Climate Change that envisions creating Paryavaran Mitra in schools. The objective of the program is to guide, facilitate, the capacity to build children through curriculum-linked activities and co-curricular actions in the club mode. Paryavaran Mitra will be groomed to take positive environmental action to have a bearing at individual, community, national and at the global level. The program is based on leveraging partnerships for Education for Sustainable Development action (CCE 2019).

Conclusion

It calls to go long miles to win the consensus of people on climate change. The WICC reports that there is a need to take action to offset some of the negative impacts of climate change on specific resources and build capacity on all kinds of stakeholders engaged in agriculture. There is a need to create a better understanding of climate science, impacts and adaptation strategies along with tools for resource managers and other decision-makers. Communicating and establishing a dialog with the public, decision-makers, community groups, local governments, nonprofits, and others about the impacts of climate change and benefits of adaptation. This will help in filling Gaps and expand knowledge about how natural and human systems will respond to climate change. The integration of climate into ongoing outreach programs of agriculture is needed. Further, there is a requirement to develop standardized climate outreach materials for climate literacy. To build the program stronger urgency to establish a channel of feedback on information and research needs.

REFERENCES

Brulle, R.J., Carmichael, J., and Jenkins, J.C. 2012. Shifting public opinion on climate change: An empirical assessment of factors influencing concern over climate change in the U.S., 2002-2010. *Climatic Change, 114*, 169–188.

Deressa, T.T., Hassan, R.M., and Ringler, C. 2011. Perception and adaptation to climate change by farmers in the Nile basin of Ethiopia. *The Journal of Agricultural Science*, 149(1): 23-31.

Climate Change Education. 2015. [*Retrieved from http://ncse.com/climate*]

Climate Change Education at CCE (2019). [*Retrieved from https://www.ceeindia.org/cee/climate-change-education.html*]

Clean Energy Workforce Development (2010). Climate Change Adaptation for State and Local Governments retrieved fromhttp://epa.gov/statelocalclimate/web podcasts/local-webcasts.html#wf1

Cooney, C. M. (2010). The perception factor: climate change gets personal. *Environmental Health Perspectives*. doi:10.1289/ehp.118-a484

Howden, S. M., Soussana, J.-F., Tubiello, F. N., Chhetri, N., Dunlop, M., and Meinke, H. (2007). Adapting agriculture to climate change. Proceedings of the National Academy of Sciences of the United States of America, 104(50), 19691–19696. doi:10.1073/pnas.0701890104

Introduction and Goals of Climate Change Education (2011).Workshop Summary retrieved from http://www.nap.edu/openbook.php?record_id=13224 and page=1

IPCC refinement guidelines of 2006 (2019) https://www.ipcc.ch/report/2019-refinement-to-the-2006-ipcc-guidelines-for-national-greenhouse-gas-inventories/

Jamelske, E., Barrett, J., and Boulter, J. (2013). Comparing climate change awareness, perceptions, and beliefs of college students in the United States and China. *Journal of Environmental Studies and Sciences*. 3(3), 269-278. doi:10.1007/s13412-013-0144-x

Lambert, J. L., and Bleicher, R. E. (2014). Improving climate change communication starting with environmental educators. *Journal of Geoscience Education*. 62(3), 388-401. doi:10.5408/13-062.1

Legesse, S. A. (2013). Farmers perceptions and adaptation to climate change in Ethiopia. IEEE Potentials, 32(5), 30-33. doi:10.1109/MPOT.2012.2222948

Leiserowitz, A. A. (2005). American risk perceptions: Is climate change dangerous? Risk Analysis, 25(6), 1433-1442. doi:10.1111/j.1540-6261.2005.00690.x

Liu, Z., Smith Jr, W. J., and Safi, A. S. (2014). Rancher and farmer perceptions of climate change in Nevada, USA. Climatic Change, 122(1), 313-327. doi:10.1007/s10584-013-0979-x

Malone, E. L. (2014). Climate perceptions. Issues in Science and Technology, 30(4), 93-93,95. Retrieved from

http://ezproxy.lib.vt.edu:8080/login?url=http://search.proquest.com/docview/1649801870?accountid=14826

McNeal, K. S., Spry, J. M., Mitra, R., and Tipton, J. L. (2014). Measuring student engagement, knowledge, and perceptions of climate change in an introductory environmental geology course. *Journal of Geoscience Education.* 62(4), 655-667. doi:10.5408/13-111.1

Nigatu, A. S., Kloos, H., Asamoah, B. O., Linköpingsuniversitet, Filosofiskafakulteten, Institutionenförtema, and Temavatteninaturochsamhälle. (2014). Knowledge and perceptions about the health impact of climate change among health sciences students in ethiopia: A cross-sectional study. BMC Public Health, 14(587)

National Climate Assessment. U.S. Global Change Research Program (2014) retrieved fromhttp://nca2014.globalchange.gov/report/response-strategies/research-needs#intro-section-2

No denying climate change. (2014). Waterloo Chronicle (Waterloo, Ontario)

Ogunlade, I., Aderinoye-Abdulwahab, S., and Mensah, A. (2014). Knowledge levels of extension agents and their perceived impact of climate change on extension service provision in Ghana. Ethiopian *Journal of Environmental Studies and Management.* 7(1), 96-103. doi:10.4314/ejesm.v7i1.12

Pahl, S., Sheppard, S., Boomsma, C., and Groves, C. (2014). Perceptions of time in relation to climate change. Wiley Interdisciplinary Reviews-Climate Change, 5(3), 375-388. doi:10.1002/wcc.272

Rejesus, R. M., Mutuc-Hensley, M., Mitchell, P. D., Coble, K. H., and Knight, T. O. (2013). U.S. agricultural producer perceptions of climate change. *Journal of Agricultural and Applied Economics,* 45(4), 701-701.

Sellmann, D. andBogner, F. X. (2012). Education in global climate change at a botanical garden: Students' perceptions and inquiry-based learning

Wisconsin initiatives on climate change impacts retrieved from http://www.wicci.wisc.edu/climate-change.php

Working Group III Contribution to AR5 (IPPCC) Climate Change 2014: Mitigation of Climate Change., retrieved fromhttp://mitigation2014.org/

Writers, S. (2014,). Farmers and scientists divided over climate change. UPI Space Daily

Yuki S., Midori A.U. (2008). Mass-media coverage, its influence on public awareness of climate-change issues, and implications for Japan's national campaign to reduce greenhouse gas emissions.

retrieved from http://www.sciencedirect.com/science/article/pii/S0959378008000964. doi:10.1016/j.gloenvcha.2008.10.005

Chapter 18

Competency Management System for Agricultural Research and Extension Scientists in India

Alok Kumar, K.H. Rao, R.V.S. Rao and P. Ramesh

Principal Scientist,
ICAR-National Academy of Agricultural Research Management,
Hyderabad, Telangana

Competency Management System (CMS) is a set of theories and processes aimed at identifying, classifying, and managing the competencies that employees need to perform specific jobs. It sets a conceptual and practical framework that drives the management of human resources to contribute efficiently and effectively to the results of an organization. This approach helps every organization's ability to maintain and enhance its human resources. This is directly linked to improving competitiveness among organizations and need to stay at the forefront of technology, starting with the development of their human capital. Competency is those abilities, required of a person in a job, which when utilized to their full potential lead to improved performance.

Competencies are an appropriate mix of knowledge, skills, and abilities of individuals in organizations that are foundations of workforce behaviors for carrying out a task. Competency Approach is widely dependent on the principle of "Right Person in the Right Job".Besides knowledge and skills, they encompass the individual's motives, attitudes, values, and other personal characteristics. Knowledge and skills are "threshold competencies" commonly used to match people to jobs, but they do not differentiate the average performer from the superior. Personal characteristics are considered "differentiating competencies", as they are significant drivers of superior performance. To operationalize, *knowledge* is a

"An appropriate mix of Knowledge, Skills and Behaviours required for carrying out a task efficiently and effectively"

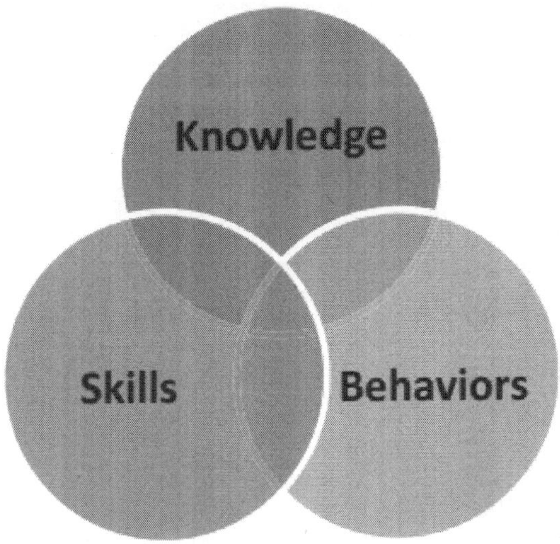

body of information, supported by professionally acceptable theory and research that individuals use to perform effectively and successfully in a given setting. *Skill* is a present, observable competence to perform a learned psychomotor act. Effective performance of skills requires the application of related knowledge and facilitates the acquisition of new knowledge. *Ability* is a present competence to perform an observable behavior or a behavior that results in observable outcomes (Lindner and Dooley, 2002). Collectively, *knowledge, skills, and abilities* are referred to as competencies. Competencies are behavioral dimensions that help to identify effective from ineffective performance (Maxine,1997). Roberts and Dyer (2004) utilized a Delphi approach to identify characteristics common across teachers that were identified as being effective in their field. This study generated a list of 40 competencies needed by agricultural science teachers that were categorized into eight areas: instruction, FFA, SAE, building community partnerships, marketing, professional growth/professionalism, program planning, and personal qualities. This report used an expert panel of agricultural science teachers, administrators, state agricultural education staff, and university teacher educators in Florida. Many years earlier, Shippy (1981) identified 246 competencies needed by beginning teachers of agricultural science teachers. He surveyed beginning and experienced agricultural science teachers, along with local agricultural education supervisors. There are several ways in which Competencies can be classified *e.g.* Core, Strategic, Functional, Technical, *etc.* Research has pointed out that as much of 65 per cent of performance problems faced in the organizations are not because people do not possess adequate knowledge or skills, but because of the differentiating competencies that go with the individual's personality and attitude which influence to a great extent to performance, output, and outcome of the employees. Over the years, the need for highly knowledgeable and skilled manpower to increase for global alignment of organizations. Human Resource Management should add value and help individuals and organizations to perform better than their present level of performance.

A competency management system (CMS) links the human resources processes to desired competencies to enable organizations to shape the capabilities of its workforce for superior performance. Many diverse and frontline global public research organizations like NASA, ESA, USDA-ARS, NIH, and others have made the strategic shift to competency-based management of human capital (NASA, 2009; USDA, 2010; NIH, 2014). NASA's CMS is an interactive computer-based Decision Support System (DSS) to (i) assess gaps between competency requirements and available competencies; (ii) provide insight into the capabilities of the workforce distributed across the programs, projects, and centers; (iii) link people with the similar competencies into communities of practice, (iv) integrate HRM with business processes and (v) search for expertise to find employees, positions or projects that utilize different competencies. The most common principle is the top-down approach. This means first identifying the broader categories and then refining them into sub-groups. In the case of scientific and technical organizations like ESA and NASA, three main levels have been identified: *Level 1:* the broad job communities, such as Engineering, Scientists, Astronauts, and Administrative Services; *Level 2:* the job families, such as Electrical Systems Engineering, Space Science, Project Management and Audit; *Level 3:* specific jobs, such as Avionics Engineer, Planetology's, Ground Segment Manager and Auditor. An additional Level 4 further details the specific technical and behavioral skills required for the given job. Once defined, Level 4 competencies may also be ranked according to proficiency levels (such as expert, master, junior) and can be used to define both the requirements of a job as well as the degree to which the person has mastered the various competencies. Competency-based management cannot rely alone on cognitive/behavioral skills but must accompany and be integrated with technical competencies. The competency aims to support the organization's goals by identifying staff with the required competencies for a given project or post, as well as identifying potential opportunities best suited to individual competencies. Knowing the capabilities of the organization's internal resources is all the more critical, now given the expected departures of a large number of experienced staff and experts shortly.

Competency-based approaches to human resources management have been effectively used by corporates and international research agencies in the past to systematically align workforce competencies with organizational goals and strategy. But their use in India is limited, particularly in institutions of National Agricultural Research and Education System (NARES), this gap in human resource management is providing to be a major bottleneck in accelerating innovation for promoting inclusive and sustainable growth of agriculture in India. Since agriculture interfaces with many other sectors of the economy, the growth of these sectors is also being constrained. An integrated, strategic competencies management system would systematically identify current and future workforce competencies and design individual competency development plans. This aligns people more effectively with institutional mission and strategy. A shift to competency-based management of human capital is also mandated by the recent Government of India's directive to implement its new National Training Policy (2012) for strategic human resources management in all its institutions. Several organizations have also developed competency dictionaries for aligning their human resources management with institutional strategy. The policy recognizes the need to introduce a competency management system that is based on every individual being viewed as a "vital resource to be valued, motivated, developed and enabled to achieve the organization's mission and objectives" (DoPT - GoI 2012).

The Department of Personnel and Training (DoPT) of the Government of India has initiated a project on "Strengthening HRM of Civil Service" in 2011 towards competency-based HRM for the Indian Civil Services. The project is being implemented to build an enabling environment for developing civil servants so that they can perform more effectively. The initiative has led to the creation of a *Competency Dictionary* and an *Implementation Toolkit* for the civil services {GoI-UNDP. (2013)}. The Indian Administrative Service (IAS) officers occupy a strategic position in the Indian public governance system. The government has generally laid a lot of emphasis on the training programmes for IAS officers to ensure that they are equipped with the right set of knowledge, skills, and attitude to effectively perform in the varied assignments that they occupy. It is increasingly being recognized that more than knowledge, skill, and attitude, proficiency in certain deep-seated abilities result in superior and more effective performance of an individual in an organizational setting. These deep-seated abilities are commonly termed as competencies. The twenty-five core competencies have been grouped in 4 sets *i.e.* Ethos, Ethics, Equity, and Efficiency, which are regarded as basic features of the Indian Civil Services and the foundations of good governance and citizen-centric administration. For moving to a competency-based approach, it would be necessary to identify the required generic competencies for the first decade of service of an IAS officer to perform their professional responsibilities effectively. The 8 core competencies as key for performing an effective District Magistrate role are: People First; Decision-Making; Planning, Coordination and Implementation; Leading Others; Integrity; Innovative Thinking; Self-Awareness and Self-Control and Problem Solving. Proficiency levels for each of these eight competencies are also important to perform excellently (Chopra *et al.*, 2014).

Competency Mapping

It is a process of identification of the competencies required to perform successfully a given job or role or a set of tasks at a given point of time. It consists of breaking a given role or job into its constituent tasks or activities and identifying the competencies (technical, managerial, behavioral, conceptual knowledge and attitude and skills, *etc.*) needed to perform the same successfully.

1. Competency Map- A competency map is a list of an individual's competencies that represent the factors most critical to success in given jobs, departments, organizations, or industries that are part of the individual's current career plan.
2. Competency Mapping- Competency mapping is a process an individual use to identify and describe competencies that are the most critical to success in a work situation or work role.
3. Competency profiling- It is the process of identifying the knowledge, skills, abilities, attitudes, and judgment required for effective performance in a particular occupation or profession. Competency profiling is a business/ company/organization/project-specific.

Steps in Competency Mapping

Competency mapping serves several purposes. It is done to resolve the following issues: Gap Analysis, Role Clarity, Succession Planning, Growth Plans, Restructuring, and Inventory of competencies for future planning. The following steps may be followed in competency mapping:

1. Decide the positions for which the competencies need to be mapped.
2. Identify the location of the positions in the organizational structure. This needs the clarity of organizational structure, defining the position relationships (reporting authority, subordinates, peers *etc.*).
3. Identify the objectives of the function or the department or the unit or section where the position is located.
4. Identify the objectives of the role. Why does the position exist? What are the main purposes of the role *etc.*?
5. Collect the Key Performance Areas of the position holder for the last two to three years from the performance appraisal records. Alternately, collect the job descriptions of any of the positions to make a list of all tasks and activities to be performed by that position holder.
6. Interview the position holder to list the tasks and activities expected to be performed by the individual. Group them into a set of tasks. The tasks list may be as many as 15 to 20 for some positions and as Competency Mapping few as five to six for other positions. There is no rigid rule about the number of tasks. It depends on how complex the position is. It is useful to start with as many tasks as possible.
7. Interview the position holder to list the actual knowledge, attitude, skills, and other competencies required for performing the task effectively. The position holder should be asked questions like: "If you are to recruit someone to perform this task what qualities or competencies would you look for in him/her? What competencies do you think are required to perform this well?
8. The process may be repeated with all the position members.

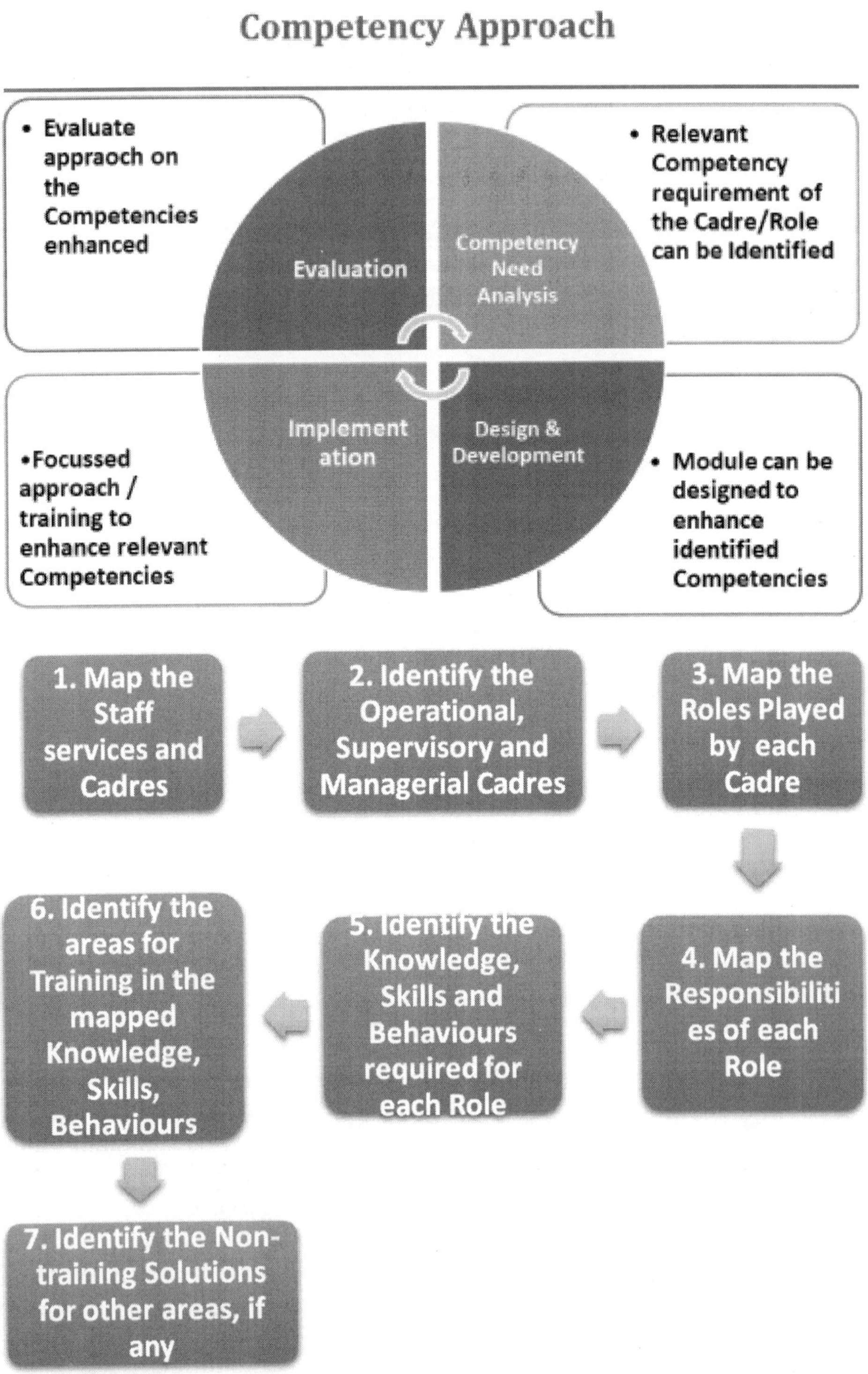

Competency Mapping Approach.

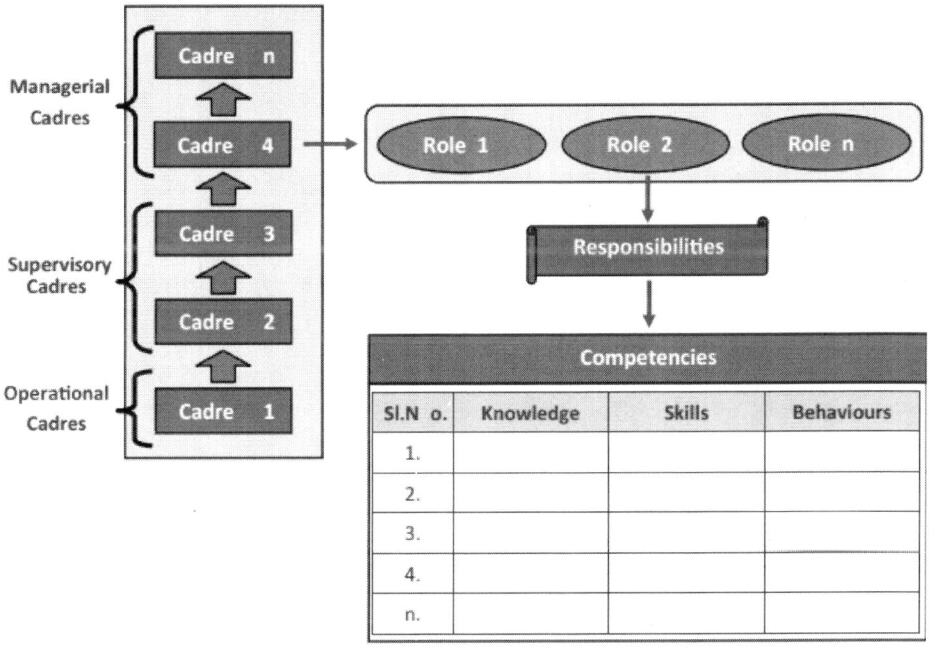

Competency Mapping Approach.

9. Consolidate the list of competencies from all the position holders by each task.
10. Edit and finalize in consultation with the position holder for their recommendation.

Indian Council of Agricultural Research and Competency Management System

The primary purpose of the Indian Council of Agricultural Research (ICAR) is to conduct research that generates knowledge to ensure a continuous flow of new technologies to increase the share of value captured by the Indian farmer. This is to be done while meeting national policy objectives of food and nutrition security, environmental security and inclusive growth of the economy. Food and nutrition security requirements of the future will be determined not only by the rising populations but also by the rapidly changing income structure of the population. The middle class, which constitutes less than a fifth of the present population will constitute about 2/3 of the population of 1.6 billion by 2050. This rapid growth of the middle class in a market economy will present a formidable challenge in assessing and meeting the demand for food and other agricultural commodities as the consumption basket changes rapidly in response to rising incomes. At the same time achieving inclusive growth in an economy, which will have a significant share of the rural population, is possible only through value addition in agriculture and increasing the share of consumer value captured by the farmer. Further, food security and inclusive growth are to be achieved while ensuring the sustainability of the natural resource base of agriculture for future generations. These challenges are further compounded by climate change. The strategy which has served ICAR so well in the relatively slow-changing eternal environment up to now is not adequate for addressing the emerging challenges and of the rapidly changing external environment of the future.

The operating environment of agricultural research is also becoming increasingly dynamic with the increasing role of the private sector in agricultural research. The rapid advances in the new sciences, molecular biology, nanotechnology, computer science, measurement technologies, mechanization, ecosystems science, simulation, big data analytics and their convergence is presenting new challenges to the organization of the agricultural science effort. Many of these new developments are proprietary to the private sector. Another emerging feature of agricultural development is the increasing role of knowledge as a significant source of value in agricultural production. New methods of knowledge generation and integration made possible through the convergence of the new sciences and new methods of knowledge transfer are being rapidly developed and deployed by the private sector. All of these imply that efficient use of resources and the rapid generation and transfer of new technologies will be possible in the future only through complementary partnerships between the public and private sectors.

The portfolio of Missions of institutions of ICAR has undergone dramatic shifts in the last decade to keep pace with the changing requirements of agricultural research and its relationships with the farmer, environment, markets, and industry. Competencies based management human resources is emerging as a key strategic tool to integrate human capital development with institutional strategy. It focuses on identifying gaps in current and future workforce competencies to meet the requirements of new institutional missions, strategies, programs, projects, and devising systems to address them. Competency-based approaches enable ICAR to institutionalize human capital development and create a vibrant organizational culture, which supports innovation and continuous learning. Eventually, a competency-based approach may impact the entire range of Human Resource Management functions: recruitment, job descriptions and placements, career planning and succession management, training and development, performance management, compensation, and benefits. With the right performance indicators and targets we could see a dramatic shift towards better service delivery as individual competencies would be aligned with organizational performance targets, breaking them down into realistic and achievable results. The National Training Policy 2012 led the way by including competency frameworks as a cornerstone for future training and development efforts by the Ministries/Departments of the Government of India.

The National Agricultural Research and Education System (NARES) comprises 114 ICAR-institutes, 73 Agricultural Universities and also supported by 714 *Krishi Vigyan Kendras* (KVKs) spread across the country. ICAR has set in motion national projects, from time to time, to assess the future human capital requirements in agriculture and allied sciences in advance to put the development of agricultural human capacity development on a rational footing. Under NAIP, a project was sanctioned to National Academy for Agricultural Research Management (NAARM), Hyderabad and Institute of Applied Manpower Research (IAMR), New Delhi on "Assessment of Future Human Capital Requirements in Agriculture and allied sectors". In the said project, the researchers assessed the demand emanating from various sectors of employment and supply from educational institutions and projected future demand and supply needs. Based on these, certain policy initiatives, strategies, and mechanisms were recommended. The forecasts provided insight into the right quantity and quality of the human resources required to maintain desired sector growth. (Rama Rao D *et al.*, 2011). The Indian Council of Agricultural Research follows the Government systems of personnel administration that continue to focus largely on the rules and procedures governing the recruitment and career development of its employees. Systematic training of scientists has continued to be mainly for the newly recruited scientists with a large number of other employees receiving sporadic training, if any at all. For transforming the Agricultural Research Services (ARS) and other employees of ICAR, it is imperative to move to a strategic human resource management system, which will look at the individual as a vital resource to be valued, motivated, developed and enabled to achieve the Organization's mission and objectives. Within this transformational process, it is essential to match individuals' competencies with the jobs they have to do and bridge competency gaps for current and future roles.

To stay competitive and relevant in intensely dynamic environments characterized by discontinuities, innovation, and institutional uncertainties, institutions need to develop strategic flexibility to adapt to unprecedented changes. Researchers have to understand the institution's strategic type and orientation to choose the set of HR practices. Specifically, some have advocated that organizations need to develop HR practices that are flexible and innovative to adapt to changing environmental contingencies. In the past three decades in China's economic reform, Chinese firms have gradually shifted away from the traditional personnel administration to innovative HR practices. Such innovative HR practices include Training and Development, Reward and Recognition, performance management, Leadership support, Peer support, Communication, Motivation, Critical Roles, Talent management, Organizational Design, Organizational culture/environment, Change management, Evaluation, Promotion, Participation in the decision-making process that are closely associated with human resource outcomes and Institutional performance. It is well known that human resource planning (HRP) is the continuous process of systematic planning ahead to achieve optimum use of an organization's most valuable assets *i.e.* quality employees. It ensures the best fit between employees and jobs while avoiding manpower shortages or surpluses. It is important to an organization like ICAR because it develops the strategies for matching the size and skills of the workforce organizational needs.

With the advancement of science and new emerging fields, nowadays, new agricultural technologies *viz.* Robotics, IoT, use of drone and sensors, precision agriculture, site-specific nutrient management, micro-irrigation, greenhouse, poly house cultivation, Drip fertigation, IDM, Hydroponics, Aeroponics, vertical farming, rooftop farming, solar farming, nanotechnology, PHT, value addition, value chain management, *etc.* have come out with distinct advantages. To feed every mouth, it is very essential to have a thorough understanding of these sciences and technologies developed in India and abroad and their application in the Indian context. For this, competent scientific manpower is the need of the hour for its proper validation and execution in the field. As per the Annual Report (2018-19) of DARE/ICAR, the total sanctioned posts of scientific cadre are 6545 and 23.1 per cent of sanctioned posts of scientific categories are lying vacant in ICAR. With the reconstructing of ASRB, it is presumed that the vacant positions particularly the Scientist staff shall be filled very soon. The current Scientific cadre ratio in position at ICAR is around 1:3.7:15.8 and it needs to be standardized. The research institution has to required its own Scientific cadre ratio to feed the need of Stakeholders of farming Community. In view of the above, it would be apt to also assess the competency level of human resources needs in ICAR to meet the national goals and objectives.

Scientific Cadre in Position at ICAR

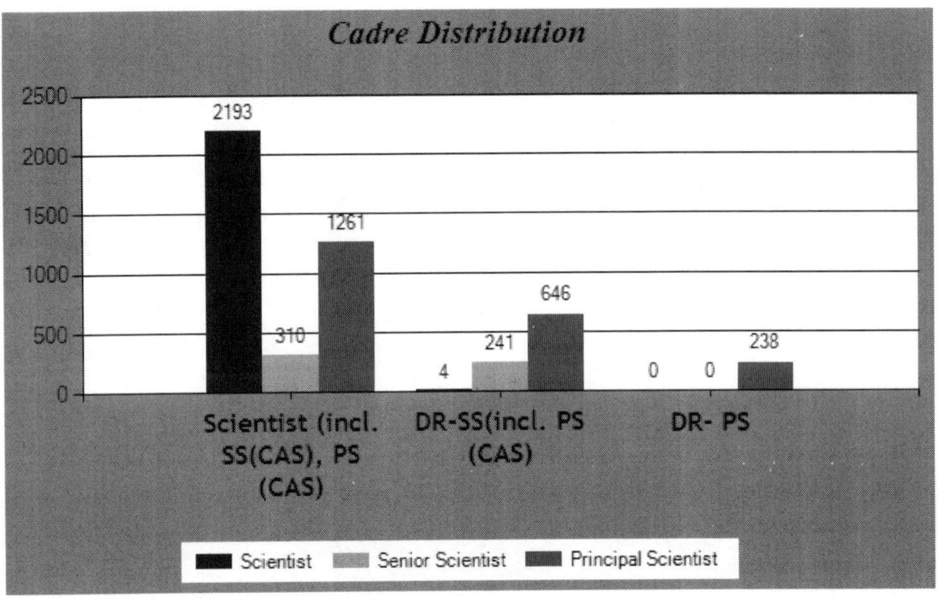

***Source*: https://pms.icar.gov.in/ as on 23.10.19**

The National Academy of Agricultural Research Management, Hyderabad has also undertaken a research project on "Assessment and development of competency framework for Agricultural Research and Extension Scientists" intending to identify the functional and behavioral competencies of scientific cadre for agricultural research and extension; and to develop a competency framework for the scientific cadre of agricultural research and extension institutions. In accordance, the identified competencies for agricultural and extension scientists are:

A) Functional Competencies and their Attributes

a) Discipline orientation (Understanding, professional, Sharing, exchange, ability, analytic, adopt, command and updating on Subject matter knowledge, Functional/technical skills, Priority setting in domain area)

b) Research attributes {Cognitive} (Ideation, creativity, Breakthrough thinking, lateral thinking, Perception, Attention, Learning, flexibility, anticipation, strategic ability for Discoveries/Invention/Innovation/ Research)

c) Research Methodologies (Designing experiment, Analytical and Statistical tools techniques, Interpretation skills on relevant data by using Information Technology)

d) Project Management and Governance skills (Execution of Plan/Operation, administration, Finance, monitoring, reviewing, reporting and Appraisal)

e) Research Communication skills {Publications (Research papers, books, book chapters, articles, reports, broachers, monographs, manuals, newsletter, leaflets, short communications, bulletins, policy papers, review papers), Impact factor, Global factor, Rating agencies} and

f) Research Application and Extension attributes (Extension, Communication, Commercialization of research, dealing with ambiguity, problem-solving, Business acumen/models, social relevance/audit, Output, outcomes and impact) and

B) Behavioral Competencies and their Attributes

a) Leading and Managing relationship attributes (Self-development, Planning, Leading, Commanding, managing, motivating, team building, decision making, time management, communication, mentoring, staffing, organizational ability, vision and purpose) and

b) Networking (Liaison, Collaboration, Linkages, Peer relationship, client-specific, Integrity, Trust building and conflict management)

These criteria were used with their relevancy, importance and level to establish the Competency Framework for Agricultural Research and Extension Scientists. Competency domains ranked at different levels of scientific cadre, Subject orientation got the prioritize for scientist level cadre indicating the development of core strength in the area of specialization, Research communication, and research methodology domains are given top priority at senior scientist cadre and principal scientist level, and at pre-Research Management Position (RMP) level, project management and governance identified as priority domain whereas Extension scientists gave the prioritize to research communication domain.

Conclusion

Competencies are deep-seated knowledge, skills, and abilities of individuals in organizations that are foundations of workforce behaviors. Besides knowledge and skills, they encompass the individual's motives, attitudes, values, and other personal characteristics. Knowledge and skills are "threshold competencies" commonly used to match people to jobs, but they do not differentiate the average performer from the superior. Personal characteristics are considered "differentiating competencies", as they are significant drivers of superior performance. Competency-based approaches to human resources management have been effectively used by corporates and international research agencies in the past to systematically align workforce competencies with organizational goals and strategy. But their use in India is limited, particularly in institutions of the National Agricultural Research and Education System (NARES). This gap in human resource management strategy is providing to be a major bottleneck in accelerating innovation for promoting inclusive and sustainable growth of agriculture in India. Since agriculture interfaces with many other sectors of the economy, the growth of these sectors is also being constrained. The study revealed that functional competencies *viz.* Discipline orientation, Research attributes (Cognitive), Research Methodologies, Project Management and Governance skills, Research Communication skills and Research Application and Extension attributes and behavioral competencies *viz.* Leading and Managing relationship attributes and Networking criteria were used with their relevancy, importance and level to establish the competency framework for agricultural research and extension scientists". A competency management system (CMS) links the human resources processes to desired competencies to enable organizations to shape the capabilities of its workforce for superior performance.

REFERENCES

Chopra, S., Kakani, R.K., and Gupta V. 2014. Preparing IAS Officer Trainees for the Role of District Magistrate: A Competency-based Approach, Research and Publication, IIM Ahmedabad,W.P. No.2014-07-02, 1-46.

DARE/ICAR Annual Report 2018-19

DoPT-GOI. 2012. National Training Policy

GoI-UNDP. 2013. Tool-Kit and Competency Dictionary for Civil Services Officers. Ministry of Personnel, Public Grievances and Pensions, 57pp; [*Retrieved from http://persmin.gov.in/otraining/CompetencyDictionary.asp*].

Lindner, J.R., Dooley, K.E., and Murphy, T.M. 2001. Discrepancies in competencies between doctoral students on-campus and at a distance. *American Journal of Distance Education,* 15(2): 25-40.

Maxine, D. 1997. Are competency models a waste? *Training and Development,* 51(10), 46-49.

NASA. 2009. NASA Competency Management System: Workforce Competency Dictionary, CMS-DOC-01, 115 pp.

NIH. 2014. Competency Dictionary, [*Retrieved from http://hr.od.nih.gov/workingatnih/competencies/core/*].

Roberts, T.G., and Dyer, J.E. 2004. Characteristics of effective agriculture teachers. *Journal of Agricultural Education,* 45(4): 82-95.

Rama Rao, D., Agrawal, R., Nanda, S.K., Awasthi, I.C., Joshi, G.P., Bhattacharya, S. and Indra Kumar, D. 2011. Assessment of future human capital requirements in agriculture and allied sectors. NAIP Project Report, ICAR-NAARM, Hyderabad and IAMR, Delhi, India. 370 pp.

Rao, R.V.S., Ramesh, P., Rao, K.H. and Kumar, A. 2017. Training Manual: Training programme on competency enhancement programme for effective implementation of training functions by HRD nodal officer of ICAR, Training manual, ICAR-NAARM, Hyderabad 103 pp.

Shippy, R.D. 1981. Professional competencies needed by beginning teachers of agriculture/agribusiness. *Journal of the American Association of Teacher Educators in Agriculture,* 22(1): 29-34.

USDA. 2010. Human capital Plan 2007-10; Taking Stock of our Human resources, 22pp.

Chapter 19

Ensuring Food and Nutritional Security through Livestock Farming

Pragya Bhadauria, Arvind Kumar and Rajbir Singh

ICAR-Agricultural Technology Application Research Institute, Ludhiana, Punjab

Introduction

India is an agriculture based country and more than 70 per cent of the farmers are landless and marginal (Kozel and Parker 2003; Taneja and Birthal 2004) where per capita land holding is hardly 0.2 Ha. In the prevailing socio-economic conditions, tradition crop farming is a serious concern for livelihood and food security in rural areas and for the entire country. Though India was successful in achieving self-sufficiency by increasing its food production and also improved its capacity to cope with year-to-year fluctuations in food production, it could not solve the problem of chronic household food insecurity. Nevertheless, the agricultural system of our country is predominantly a mixed crop-livestock farming system in which crop production is combined with the rearing of livestock, which supplements the farm incomes. Therefore, the future of sustainable agriculture growth and food security in India depends on the performance of allied sector especially on livestock sector. According to NSSO estimates in 2014, agriculture sector contributes to around half of the income of farmers and more than one-tenth is being contributed by livestock sector, thus making contribution of agriculture and allied sector more than 60 per cent in total income of the farmer. The small and marginal farmers have only 44 per cent of the land, but contribute to more than 80 per cent of livestock and fishery. The conducive climatic and topographical conditions of India led to emergence of livestock sector as a key socio-economic driver of rural households. The small farmers, especially the landless livestock holders, are extensively dependent on income from the sale of milk and animals to meet their daily household expenses. They not only contribute to their income, but food energy intake at household level is now given prominence in assessing food security. Thus, the livestock sector has emerged as one of the important drivers of agricultural growth and diversification in India. The objective of this paper is to examine the role of various livestock farming, its prospects and challenges in achieving food and nutritional security in the country.

Background

Mankind has been utilizing different animal species from the dawn of civilization for a variety of purposes *viz.* production of milk, meat, wool, egg and skin/leather. Apart from these, various animal species are also used for draught power, companionship, entertainment, research experimentation, sports, security *etc.* However, livestock includes domesticated animals intentionally reared in an agricultural setting for want of food or fibre, or for their labour. It includes Pigs, Cattle, Goats, Sheep, Poultry, Horses, Donkey, Camel, Yak, Mithun *etc.* The type of livestock reared varies region-wise and depends on factors such as climate, consumer demand, native animals, local traditions, and land type. India has a rich heritage of rearing livestock and animal husbandry is culturally and economically imbibed into the society. Thus, they have been responsible for developing and conserving domestic animal diversity with important genetic traits. Traditionally, cattle rearing in India are not just confined to milk production alone, it has religious sentiments too. Animals like oxen, male buffaloes, ponies or mules are utilised for draught power as a cheap modes of transport for the poor farmers. Goat and sheep are known as the poor man's cow or bank on hooves which survive with least resources. In this way livestock plays a multifaceted role as food source (by providing nutrient-rich food products), generate income and employment and act as a cushion against crop failure, provide draught power and manure inputs to the crop subsector, and contribute to foreign exchange through exports (Birthal and Rao, 2002).

Nutritional Importance of Animal Source Foods

The diet of an average Indian is cereal-based and lacks nutrient-rich foods such as pulses, fruits, vegetables, and animal products. Low intake of these products results in nutritional deficiencies. About 30 per cent of the Indian population suffers from malnutrition (Kumar and Joshi 1999). A more sustainable approach for subsistence farming households is diversification of their food production through the introduction of animal based diets that are suited to local agro-ecological conditions and can fill macro- or micronutrient deficiencies in the local diet. Diets based on animal products contain proteins, carbohydrates, fats, minerals and vitamins present in appropriate amounts and readily digestible forms to meet all human nutritional requirements. Consumption of these products being protein, vitamin and micronutrient rich is expected to have positive impact on nutrition and health, especially of the poor. Therefore, livestock directly contribute to nutrition security in the form of *"animal-source foods"* by converting low-value materials, inedible or unpalatable for people, into milk, meat, eggs and fish. Protein-energy malnutrition, iron-deficiency anaemia, and vitamin- A deficiency, all of which can be prevented if sufficient animal-source foods are included in the diet (WHO, 2009). Animal products also contribute one-third of the total protein consumed by the people. It is an important part of the diet because it functions as a building block for bones, muscles, cartilage, skin and blood. By providing essential nutrients, especially in the first critical 1,000 days from conception, animal-source foods can help ensure normal physical and cognitive development. Protein also helps the body produce the enzymes, hormones and vitamins it needs. Although, in India, only about 10g of protein is derived daily per capita from livestock products compared to 16.6 g in Asia in general, 45.3 g in the Americas, 47.7 g and in the European Union (FAO, 2011).

Animal products, such as dairy products from cows, sheep and goats served as significant source of both energy and protein, including many essential amino acids and micro-nutrients such as Vitamins A and D. Cheese and yogurt are other dairy products that provide calcium, potassium, vitamins and protein that help the body grow and develop. Thus, increased dairy products consumption is therefore assumed to improve nutritional outcomes for households. Other animal products such as meat of poultry, beef, pork and fish, supplies many nutrients. These include protein, B vitamins (niacin, thiamin, riboflavin, and B6), vitamin E, iron, zinc and magnesium. Furthermore, chicken is cheaper than other protein sources and it does not have any social taboo. Other meats such as chevon, mutton or fish are available, but highly priced. Eggs also supply many of these important nutrients which is the cheap and best animal protein sources for human beings. Nevertheless, among the various livestock species, piggery has the most potential source for meat production as it is the most efficient feed converters after the broiler. Pork acts as important source of cheap animal protein for the weaker sections who can-not afford costlier meat.

However, in Indian context, culture, traditions, customs, and taboos influence animal product consumption to a great extent. Albeit, studies show that urbanization has been causing a rise in demand for meat products. On account of favourable socio-economic factors such as changing eating habits, higher purchasing power, urbanization, increasing health consciousness towards protein rich diet, there has been increase in demand for meat and the sector has gained importance in terms of contribution to income, employment and foreign exchange. Milk is the main output of the livestock sector, providing 67 per cent of the sector's total value, while meat and eggs provide 17.5 per cent and 3.6 per cent, respectively. Consumption of livestock products has been increasing both in rural and urban areas and livestock represents nearly 22 and 25 per cent respectively of the food basket (NSSO survey, 2009-10). Milk consumption increased by 70 percent and meat consumption by 30 percent, with the latter largely driven by the rise in poultry consumption. Thus, expenditure on milk increased at a higher extent than that on meat. Similarly, the expenditure elasticity of demand for dairy and poultry products have been estimated by various studies to be 0.99–1.32 for urban areas and 1.15–1.96 for rural areas (Sharma *et al.*, 2008), this means that demand growth is expected to be higher in the rural areas affording livestock to be a source of sustained and inclusive growth. Thus, increased expenditure on livestock products by poor people in the rural areas potentially contributes to improved nutrition and health.

Role of Different Livestock Sectors in Food and Nutritional Security

According to World Food Summit in 1996, food security is defined as "when *all people, at all times, have physical and economic access to sufficient, safe, and nutritious food that meets their dietary needs and food preferences for an active and healthy life*". People can be malnourished if they eat insufficient calories or protein, food of poor quality lacking sufficient protein and minerals, or if they are unable to utilize fully the food they eat. According to World Bank (2015), 30 per cent of Indian population lives below poverty line and more than 70 per cent of these reside in rural areas with meagre means of nutrition and hence food insecurity. These masses can be targeted and brought under the umbrella of food security by promoting livestock farming amongst them and helping them to sustain their food demands. The contribution of livestock to food, distinguished from nutrition security among the poor, is mostly indirect: sales of animals or produce, demand for which is rapidly growing, can provide cash for the purchase of staple foods, and provision of manure, draft power, and income for purchase of farm inputs can boost sustainable crop production in mixed crop-livestock systems.

Dairy Sector

Among various livestock products, milk constitutes the major share (67 per cent) in value of outputs from the animal husbandry sector and is the single largest commodity contributing to the value of output from agriculture. India's milk production has increased up to 176.35 MMT in 2017-18, with a growth rate of 6.62 per cent and the per capita availability of milk in India during 2017-18 was 375 gm/day (DADF, 2017-18). With this the country ranks first in global milk production. It provides livelihood and nutrition to millions of homes in villages, ensuring supply of quality milk and milk products to people in both urban and rural areas. As most of the dairy farmers in India are raising animals at small scale by traditional method that are playing an important role in the total milk production and economy of the country. Women accounted for 93 per cent of total employment in dairy production. Traditionally milk is consumed in liquid form (raw but usually boiled, and pasteurized) and in the form of various locally processed products *e.g.* Ghee, Makkhan, Dahi, Paneer, Lassi, Khoa, Rabdi, Channa, Barfi and other sweets can be made easily from the milk and can be sold in the market at better prices and can also be consumed by the family. Milk consumption levels were higher in urban areas, but now shown an increasing trend in rural areas also. The sector needs focused attention particularly in drought prone area where there is all the more need to add to the incomes of the farmers. Farmers can be trained to switch to scientific dairy practices by procuring elite dairy animals and better husbandry practices for increasing the milk production that will further enhance their livelihood and serve as complete food for others.

Poultry Sector

Within the livestock sector, poultry has been the fastest growing sub-sector. The sector has grown at a rate of 6 per cent over the years. India is currently ranked

3rd in egg production and 4th in chicken meat production in the world. Poultry sector is broadly divided into two sub-sectors – one with a highly organized commercial sector with about 80 per cent of the total market share and the other being unorganized. The commercial production systems are expanding at a fast pace while the household poultry remains a major source of supply for local consumption and meeting nutritional needs of rural poor. This unorganized sector also referred to as backyard poultry plays a key role in supplementary income generation and family nutrition to the poorest of the poor. It serves as a backbone for the *"Food Security Model"* for the rural poor. Village poultry have constantly commanded a price premium over commercial birds and there is a wide market demand for village poultry products. The government has supported many projects which led to the evolution of this sector among the rural areas. Backyard Poultry rearing has significant advantages over other livestock activities because it can be taken up under diversified agro-climatic conditions with minimal investment and can be easily managed by the children and ladies of the house. Rearing chickens can be the easiest way of obtaining daily nutritional requirements especially to the women and their children. Egg is complete diet, except vitamin-C and fibre and it is the only food which is unadulterated with high biological value. It is the cheapest animal protein having more shelf life and has no social taboos. Eggs supply an array of vitamins such as A and B12, and K, choline that plays a role in brain development. Chicken and egg consumption increased considerably in both rural and urban areas due to health consciousness.

Small Ruminants Sector

The small ruminant sector, comprising of sheep and goats, constitute an important role in the agrarian economy, especially in areas where crop and dairy farming are not economical. Unlike larger stock such as cattle, small livestock require less space; they are less capital intensive and are easy to manage as they can be reared within or near homesteads. This makes it much easier for women who look after the small livestock along with routine household activities. The rearing of small livestock near or within homesteads ensures regular supply of food to the families in terms of meat and milk. Sale of livestock and their products can be a valuable source of income. For example, animals, especially small livestock can be sold to meet immediate family needs such as food, clothing and other expenses. In many nomadic societies and small mixed crop livestock systems, goats and sheep are the major source of milk for the family even though production per individual animal is small. They have short generation intervals, higher rates of prolificacy, and marketing of mutton and chevon products is far easier. Some people are intolerant of lactose, especially in bovine milk; goat milk is often a substitute for cow milk in these cases and is well tolerated by almost everyone. Goat milk has therapeutic properties and is easily digestible due to smaller size of fat globules and serves as a ready source of family nutrition. It is a rich source of nutritionally available minerals and it contains more of calcium and phosphorus than cow and human milk. Goats are among the main meat-producing animals in India, whose meat (chevon) is one of the choicest meats and has huge domestic demand especially during religious functions.

Piggery Sector

According to the 19th Livestock census of India pigs comprises 2.01 per cent to the total livestock population. Pig rearing has the potential to attain stability of nutritional status at a comparatively low expenditure and a sustainable source of income. Pig rearing practices are still very traditional in nature and are mostly being done by backward and tribal population of the society. Cost of inputs and returns are not a serious concern in this low or zero input pig rearing system. It is also one of the most efficient food converting animals among domesticated livestock and can play an important role in improving the nutritional status and ensure food security especially for weaker sections of the society. The pork provides high amount of animal fat which is a rich source of energy as well as other important micronutrients and macronutrients. The lard present in pig meat imparts a typical flavour to the cooked pork, which has made it popular among a large mass of people especially in north east region of India.

Indirect Effects of Livestock on Food Security

Livestock generate cash incomes from the sales of animals, their products, or services, or through employment along with animal-source food value chains. In addition to directly providing cash-generating opportunities for livestock keepers, also provides livelihood to two-third of rural community and contributed 16 per cent to the income of small farm households as against an average of 14 per cent for all rural households (DADF, 2017). The increase in demand for livestock products witnessed over the past few decades has created significant opportunities for small farmers who can raise animals to meet that demand and get benefit out of this. Livestock value chains represent a large and growing employment sector. They include farm-level production, input, and service industries to farmers; transportation of livestock and their products; and processing and marketing. Livestock indirectly contribute to food production by providing manure, contributing to land preparation, and providing ready cash to buy planting materials or fertilizer or to hire labor for planting, weeding, or harvesting. Although on aggregate, draft power use is decreasing globally, but in some regions of India, it continues to contribute significantly to food production.

Issues and Challenges

Although, poor and malnourished are more likely to benefit from the consumption of livestock products, but without livestock ownership their consumption generally remain very low. Due to income constraints, they are least likely to consume such products or adequately consume to derive any significant benefit out of them. In case, if they posses animals their produce such as milk and eggs were more likely to sell all or a portion of the output rather than consuming at home in order to buy grains and other necessities in which they may have deficit. In such cases livestock ownership may not help alleviation of under nutrition in the short run. Livestock can also impact food security by transmitting diseases to people via vectors such as biting flies and through contaminated animal-source foods; these diseases limit productivity of people by reducing their ability to produce food themselves or to work to earn income to purchase food. Similarly, overconsumption of animal-source foods such as fatty red meats and hard cheeses can harm human health; can lead to cardiovascular disease, while overconsumption of processed meats, such as bacon and ham, has been associated with some cancers. Increased consumption of energy-dense meat, milk, and eggs also contributes to the obesity.

Conclusion

Agriculture and animal husbandry have been a part of Indian economy since long and contribute significantly in rural and economic development of the country. Over the coming decades, population growth, urbanization, and income growth, especially in developing countries, will result in huge increases in demand for milk, meat, and eggs. Livestock based foods are complete in every aspect being rich in energy, protein and lipids essential for overall development of the human body. Pigs and poultry are likely to remain the main source of meat where rapid urbanization is occurring. Sale of value added products paves way for an extra income to the farmer to sustain their economy. Demand for animal products is set to continue increasing in the next three decades, as is their market price. Therefore, in agriculture based economy real development can be achieved only by integrating livestock and agriculture for long term sustainability goals.

REFERENCES

BAHS, GOI. 2014. Basic Animal Husbandry Statistics. Department of Animal Husbandry, Dairying and Fisheries. Ministry of Agriculture. New Delhi.

Birthal, P.S. and Rao, P. 2002. Technology Options for Sustainable Livestock Production in India. Proceedings of the Workshop on Documentation, Adoption, and Impact of Livestock Technologies in India. ICRISAT Patancheru, India.

Birthal, P.S. and Ali, J. 2005. Potential of Livestock Sector in Rural Transformation, In: Rural Transformation in India: The Role of Non-farm Sector (Rohini Nayyar and A N Sharma editors) Institute for Human Development and Manohar Publishers and Distributors, New Delhi.

Department of Animal Husbandry, Dairying and Fisheries (DADF) Annual Report 2017-18, Ministry of Agriculture and Farmers Welfare, Govt. of India.

FAO, 2008. Poultry Sector Country Review, India version. [*Retrieved form http://www.fao.org/3/a-ak069e.pdf*].

FAO, 2011. World Livestock 2011- Livestock in Food Security. FAO, Rome.

Karmakar, K.G. and Banerjee, G.D. 2006. Opportunities and Challenges in the Indian Dairy Industry.

Kozel, V. and Parker, B. 2003. A Profile and Diagnostic of Poverty in Uttar Pradesh, *Economic and Political Weekly*, 38(4): 385-403.

Kumar, P. and Joshi, P.K. 1999. Socioeconomic factors contributing to nutritional security of rural households. In Diversification of agriculture for human nutrition. National Academy of Agricultural Science and National Institute of Nutrition, Hyderabad.

19th Livestock Census-2012. All India Report. Department of Animal Husbandry, Dairying and Fisheries. Ministry of Agriculture. New Delhi.

Randolph, T.F., Schelling E., Grace D., Nicholson C.F., Leroy J.L., Cole D.C., Demment M.W., Omore, A., Zinsstag, J. and Ruel, M. 2007. Role of livestock in human nutrition and health for poverty reduction in developing countries. *J. Anim. Sci.*, 85, 2788–2800.

Sharma, V.P., Nin Pratt, A., Kumar, A., Singh, R.V., and Staal, S. 2008. Dairy development in India. In: Staal, S, Nin Pratt, A and Jabbar, M A (eds). Dairy development for the resource poor: Part 2- Pakistan and India Dairy Development Case studies. FAO. Pro-Poor Livestock Policy Initiative, Rome, Italy and ILRI Nairobi, Kenya.

Taneja, V.K. and Birthal, P.S. 2004. Role of Buffalo in Food Security in Asia, *Asian Buffalo Magazine,* 1(1): 4-13.

WHO (World Health Organization). 2009. Global health risks: Mortality and burden of disease attributable to selected major risks. WHO, Geneva.

World Bank. 2015. Ending Poverty and Hunger by 2030. An Agenda for the Global Food System.

Chapter 20

Empowering Farm Women vis-à-vis Animal Husbandry Sector

Y.S. Jadoun[1] *and Pragya Bhadauria*[2]

[1]*Assistant Professor,*
Department of Veterinary and Animal Husbandry Extension Education
Guru Angad Dev Veterinary and Animal Sciences University,
Ludhiana – 141 004, Punjab
[2]*ICAR-Agricultural Technology Application Research Institute,*
Ludhiana – 141 004, Punjab

"If you want something said, ask a man; if you want something done, ask a woman."

— *Margaret Thatcher*

India is an agriculture based country and livestock sector is an integral component of it. It is a source of food, income, employment and foreign exchange. For small income producers, it provides draught power and organic fertilizer for crop production; serves as store of wealth and means of transportation as well (Garcia, *et al.*, 2006). About 90 per cent of livestock is owned by small farmers and landless rural households. These small holders, especially the landless livestock holders, are extensively dependent on income from the sale of milk and animals to meet their daily household expenses. Even in the mixed farming systems, livestock is the main source of income for rural households and a source of productive employment for the poor.

The livestock production is largely in the hands of women and they are performing phenomenal work in the rural areas (Waters Bayers, 1985). Most of the livestock activities are incomplete without the assistance of women. Economic Survey 2017-18 says that with growing rural to urban migration by men, there is 'feminisation' of agriculture sector, with increasing number of women in multiple roles as cultivators, entrepreneurs, and labourers. In fact animal husbandry is becoming feminized and they contribute 60 to 80 per cent of labour in the animal husbandry (Younas *et al.*, 2007). They undertake various activities of livestock management such as fodder collection, grazing of animals, care of animals, feeding, watering, and health care, management, milking, cleaning of animal sheds *etc.* Their involvement in the post-harvest operations like household-level processing, value addition is the exclusive responsibilities of women. They also prepare cooking fuel by mixing dung with twigs and crop residues. So one can say rural women remain busy from dawn to dusk in various agricultural activities and livestock management.

Women constitute around 50 per cent of the global population. A large number of women around the world are unwaged. The world economy suffers a lot because of a disproportionate opportunity for women at work places. In India, concept of gender sensitization is of recent origin. Women have become aware about their rights and situations and entered in different fields of business. They have established their own successful business empires. They are contributing towards the growth of economy and improvement of their socio economic conditions. Government of India has given due importance to gender sensitization and women empowerment in the country (*Borkar et al., 2017*)

Concept of Gender Mainstreaming and Women Empowerment

Gender mainstreaming is an approach to policy-making that takes into account both women's and men's interests and concerns. The concept of gender mainstreaming was first introduced at the 1985 Nairobi World Conference on Women. Women have undergone a radical transformation from merely a homemaker to a dynamic multifaceted personality contributing to the socio-economic growth worldwide. As per NSSO, the present share of workforce in agriculture, livestock, forestry and fishery is more than 50 per cent in all states. Despite considerable involvement and contribution of women in the society, their contribution has not been duly acknowledged. Significant gender inequalities also exist in access to technologies, credit, information, inputs and services probably because of inequities in ownership of productive assets including land and livestock. The process of creating this knowledge and awareness 'of' and responsibility 'for' gender among professionals enabling them to address gender issues in appropriate and most effective ways is called 'gender mainstreaming'. Mainstreaming is not about adding a "woman's component" or even a "gender equality component" into an existing activity. It goes beyond increasing women's participation; it means bringing the experience, knowledge, and interests of women and men in any policy, programme, reform or activity or any developmental agenda.

Similarly, women empowerment refers to the creation of an environment for women where they can take decisions of their own for their personal benefits as well as society. It is an active, multidimensional process which enables women to realize their potential and powers in all spheres of life including economic empowerment. Economic empowerment is nothing but making women aware about their role/ importance in economic development and provide them space for attaining financial independence and account their significant contributions in an enterprise.

Livestock and Rural Women

In India, women in animal husbandry and livestock sector constitute 69 percent of the labor force as against 35 percent in crop farming (*Economic Survey 2002-03*). They form the backbone of agriculture and allied activities, comprising the majority of agricultural labourers and play a significant and crucial role in agriculture and livestock developmental activities. Livestock production is largely in the hands of women. In fact animal husbandry is becoming feminized Women do a bulk of livestock related activities like fodder collection, feeding, health care of animals, watering, milking and household level processing (*Planning Commission, 2012*). Women accounted for 93 per cent of total employment in dairy production (*Qureshi et al., 2016*). Depending upon the economic status, women perform the tasks Such as fodder collection, feeding, watering, and health care, management, milking and household-level processing, value addition and marketing *etc.* are performed by women. Despite of their significant role and considerable involvement and contribution in livestock management and production, women's control over livestock and its products is negligible and there is significant gender inequalities exist in access to technologies, credit, information, inputs and services probably because of inequities in ownership of productive assets including land and livestock. Therefore, there is a need to correct gender bias in livestock sector, veterinary education, research and service delivery systems as to enhance the effectiveness of women-oriented livestock development programs.

How to Develop Women Entrepreneurs through Livestock Sector?

Right efforts on from all areas are required in the development of women entrepreneurs and their greater participation in the livestock based entrepreneurial activities. Following efforts can be taken into account for effective development of women entrepreneurs in livestock sector.

- ☆ Consider women as specific target group for livestock based developmental programmers.
- ☆ Vocational training to be extended to women community that enables them to understand the production process and production management.
- ☆ Adequate training programmed on management skills to be provided to women community.
- ☆ Encourage women's participation in decision-making.
- ☆ Training on professional competence and leadership skills should be given to women entrepreneurs.
- ☆ Training and counselling on a large scale of existing livestock related women entrepreneurs to remove psychological causes like lack of self-confidence and fear of success.
- ☆ Better educational facilities and schemes should be extended to women folk from government part.

- ☆ Continuous monitoring and improvement of animal husbandry related training programmers.
- ☆ Activities in which women are trained should focus on their marketability and profitability.
- ☆ To encourage more passive women entrepreneurs the women training programs should be organized.
- ☆ Counselling through the aid of committed NGOs, psychologists, managerial experts and technical personnel should be provided to existing and emerging livestock women entrepreneurs.

Gender Mainstreaming through Livestock Entrepreneurship

Entrepreneur associated to livestock farming/business, production of livestock related raw materials and livestock related processing industries is considered as livestock entrepreneur. In other terms, a person who is linked directly or indirectly to the animal husbandry or livestock sector is referred as livestock entrepreneur.

1. Dairy Sector

Since the dairy industry has become more consumer oriented due to health consciousness and increase in purchasing power leads to development of innovative practices of organized retailing, supply chain management, balanced product portfolio, product development. It leads to increase demand for milk and milk products in recent years. Dairy farming is one of the important enterprises which dominate the economic activities of the woman in the rural areas of India. Women accounted for 93 per cent of total employment in dairy production. Depending upon the economic status, women perform the tasks of collecting fodder, watering, feeding and other animal management activities. Therefore, it serves as a profitable enterprise for rural women as they are well aware about animal behaviour and production characteristics.

Women can be guided for small scale dairy farming and commercial dairy business. They can start their own dairy farms by rearing small number of animals and with proper business plan, scientific management and care can ensure maximum production and profit from dairy farming business. If woman has to expand her dairy enterprise on a commercial basis successfully, she has to adopt new and modern dairy farming tools, time and energy saving tools/drudgery reducing technologies, *etc.* As there is more than 60 per cent fodder deficit in India, they can also start seed/fodder banks in the potential areas. Women can combine together, purchase fertile land and produce quality fodder and supply them to the nearby livestock farmers. For this they need to undergo specialized trainings from various organisations.

2. Goat Rearing

High demand for goat and its products with potential of good economic returns have been deriving many progressive farmers to take up the goat enterprise on a commercial scale. Goat is mainly reared for milk, meat, sale of breeding stock as an income source. This versatility allows the producer to plan and operate a more stable economic production unit. In many parts of the India, some of the goat breeds are raised for fiber, meat, and milk and cheese production. In general, the role of women in goat keeping is very significant in the rural families and goat is the most important means through which rural women are able to contribute meaningfully to the cash needs for her and their family members. Most of the activities like sale/ purchase of goat, grazing of goats, feeding and watering, care of kids, cleaning of pen and household maintenance, milking of goats, health management and breeding management of goats were performed mainly by women. Thus, goat rearing is the most useful way of women's earning those who stay at home. Among the livestock livelihood, goats are more economical than cattle as they are less investment oriented. An increased level of adoption of technologies and availability of good quality breeding stock would be essential to make the commercial goat farming more successful. Simultaneously, emerging favourable market conditions and easy accessibility to improved goat technologies are also making this enterprise a profitable venture for these women.

3. Poultry Sector

Poultry farming is an important livestock activity for rural women as it generates cash income and provides employment opportunities and improves household nutrition. Rural women are primarily responsible for the care and management of the bird under backyard poultry systems. Poultry rearing has significant advantages over other livestock activities to rural women, because: it is easy to manage and can be taken up under diversified agro-climatic conditions. It can provide quick returns and constant source of income throughout the year, as it enjoys good market demand and prices. Backyard Poultry Farming (BYPF) enhances women's social status and decision-making power in the household by increasing women's income and can be used as a tool to reduce poverty in rural areas. The BYPF rearing had shown a remarkable impact on women's livelihoods across the country. Since women got to deal with different people while purchasing and marketing birds or eggs, these women slowly gained confidence. Various centrally and state government sponsored schemes are promoted for empowering the BPL families mainly the women farmers *e.g.* National Livestock Mission (NLM), Rural Backyard Poultry Development (RBPD) which covers beneficiaries from BPL families to enable them to gain supplementary income and nutritional

4. Aquaculture

The Indian aquaculture is mainly male-dominated and that is largely because it involves fieldwork, travel and living in remote areas. But it has increased significantly with the emergence of fish processing as a growth area within the manufacturing sector in the past years. A recent FAO study found that women's involvement in rural aquaculture had led to a 10- 20 per cent increase in fish production and almost all women fish farmers shown a significant improvement in their socioeconomic conditions. Women play a major role in aquaculture production as labourers and managers of the production process: fishing, processing and marketing. Women form the core of the industrial fisheries labour force through their involvement in postharvest or processing activities. The involvement of women in these activities generates supplemental income to support their families. Some of the government agencies, NGOS, NABARD, NFDB have identified several fisheries technologies, which could facilitate women to become entrepreneurs and enhance their incomes. The various fields that can be chosen by women entrepreurs in this sector are mainly composite fish culture prawn culture, integrated fish culture along with horticulture and animal husbandry, backyard hatcheries, fish-feed manufacturing, fish marketing (wholesale/retail), fish processing and packaging, value addition, net mending/repairs, ornamental fish farming *etc.* Financial support to these enterprises is provided by NABARD through a rural credit system which provides capital for short-term production/marketing activities, and medium term and long-term loans for technically feasible and financially viable projects through State Cooperative Banks, State Cooperative Agriculture, and Rural Development Banks, Regional Rural Banks and Commercial Banks.

Approaches and Strategies for Gender Mainstreaming through Livestock Sector

1. Access to Financial Services and Financial Literacy

Lack of access to credit was a major constraint for poor women workers to develop any economic activity. Public banks were not adapted to these women who were mainly illiterate, had no collateral and little knowledge about how to use or manage credit efficiently. Rural Bank or cooperative-type bank are the right choice that targets poor women from the informal sector. Its main objective is to help women come out of poverty by providing them with financial services adapted to their situation and capacities. These Bank designed financial products to meet the women's needs and the bank created a door-to-door service so that savings could be collected regularly without requiring the women to travel to the bank. The services provided by the bank help women avoid exploitative moneylenders build their own savings, assets and learn about financial and business management. Along with this awareness should be created about various credit facilities, financial incentives and subsidies through various channels. Government and NGOs offered various schemes and opportunity to the rural entrepreneurs. But, they are unaware of these schemes and opportunities due to their illiteracy. So they should to be educated by the conducting workshops and seminars related to their business. Financial institutions like ICICI, SIDBI, IDBI, IFCI, and SFC should also be encouraged to provide finance to rural entrepreneurs with low rate of interest and limited collateral security with liberal terms and conditions. Simulteously, rural women could be motivated to avail finances from these kinds of institutions for starting the entrepreneurial activities.

2. Promotion of Livestock Based Self Help Groups (SHGs)

The SHGs are voluntary associations of people formed to attain some common goals. Members in SHG agree to save regularly and convert their savings into a common fund and to use this common fund for management and business activities. The biggest problem with the SHGs is increasing substantially in rural areas as majority of them are unorganized. These groups are mobilizing thrift deposits, but unable to receive timely, matching and revolving funds to generate employment activities to earn their livelihood. To overcome these constrains SHGs are being actively promoted by government of India through Ministry of Rural Development and Employment, RBI, NABARD, DRDA, Line Departments of State Governments, NGOs, *etc.* at various capacities. NABARD has been working as a catalyst in promoting and linking more and more SHGs to the banking system. Simultaneously the SHG should be promoted to adopt latest technologies to thrive in various market and that could be only possible if they are provided with extension and training support on a sustained manner. Therefore, there is a need for the extension workers, researchers and technologists to provide a helping hand to them.

3. Establishing Milk Cooperative Societies for Women at Village Level

Women dairy Cooperatives (WDC) may serve as good option for rural women to become more empower as they are authorised to make their own decision outside the home. Most of the women dairy co-operatives in India are based on the principle of maximization of farmer profit and productivity through cooperative effort. This pattern, known as the Anand Pattern, is an integrated cooperative structure that procures, processes, and markets produce. (Datta and Ganguly, 2002). The dairy activity is now largely based upon a three tier system under which, the primary village cooperative societies of women are linked with district union and state federation which are guided by the national co-operative dairy federation in India. For *e.g.* Gujarat Co-Operative Milk Marketing Federation – AMUL, Karnataka Milk Federation – NANDINI, Rajasthan Co-Operative Dairy Federation– SARAS, Bihar Milk Federation– SUDHA, Tamilnadu Co-operative Milk Producers' Federation Limited–Aavin.

4. Women based Producer's Organizations

A Producer Organisation (PO) is a legal entity formed by primary producers, *viz.* farmers, milk producers, fishermen, weavers *etc.* A PO can be a producer company, a cooperative society or any other legal form which provides sharing of benefits among the members. As small producers are not able to harness the benefit of economies of scale. Besides in livestock marketing, there is a long chain of intermediaries who very often work non-transparently leading to the situation where the producer receives only a small part of the value that the ultimate consumer pays. Therefore, through aggregation the groups of women producer can avail the benefit of economies of scale. They will also have better bargaining power vis-à-vis the bulk buyers of produce and bulk suppliers of inputs. Promoting Livestock based producers organization run by women ensures investment of various stake holders involved in animal husbandry sector to improve the production and profitability. Creation of a separate label for women's products which can give women greater recognition for their work, and develop a sense of pride and ownership of their products. Similarly, Promotion of value addition to the livestock products such as milk, egg, meat, and fish has huge profit potential. Value of the products get increased many folds during processing, and thereby provide excellent returns. Women can start their own milk parlour, where they can sell processed milk and milk products like flavoured milk, khoa, ice cream, paneer, shrikhnad, curd, gulabjamun, whey milk, ghee *etc.* or meat centre where fried chicken, chicken, mutton keema, *etc.* could be sold. Marketing of these values added products could be done in their own brand name and they can start chain of parlours/hotels thereafter. Women may also opt for converting farm wastes (dung, urine, wasted fodder *etc.*) especially of local or deshi cows to produce biogas, vermicompost plant elixirs and pesticides *etc.* and sold as products that have higher market values. NABARD provides financial support to these POs through "Producers Organisation Development Fund". Government of India provides budgetary support to these POs for its Equity Grant and Credit Guarantee Fund Scheme for the Farmer Producer Company. For creation of storage and other agricultural marketing infrastructure under the Integrated Scheme for Agricultural Marketing (Ministry of Agriculture, Government of India), FPOs are eligible to get higher subsidies. CAPART, Ministry of Rural Development also operates schemes through which support for some activities can be obtained by the PO.

5. Promotion of Organized Farm and Dairy/Meat/Egg Market

Understanding the market and preparing it to respond to emerging market trends would be the prime instrument for enhancing the women livelihood opportunities in the livestock sector. Rural women entrepreneurs are facing tough competition from the large scale organizations and urban entrepreneurs. The women producers are not collective in their approach for marketing their products because they are to widely scattered and mostly uneducated. The women entrepreneurs are heavily dependent on middlemen for marketing of their products who pocket large amount of profit. Government should take steps to provide market information of different products and finance to the rural entrepreneurs at concessional rates. They should be encouraged to have a regular income by providing market-related and technical support and identifying local and national marketing opportunities. Innovative technology should be publicly announced in order to get its acceptance among the rural entrepreneurs. Assess to infrastructural and storage facilities and means of transport are other means of marketing to be supported to these women in rural areas. Integrated marketing services should be provided to the rural producers through exploring, developing and establishing first the internal (local market) and then the national market for their products. Various models like rural haat, trade fairs, livestock fairs *etc.* facilitate linkages with markets and private companies should be promoted for showcasing their products. The promotion, protection and branding of women products is an important strategy to increase female participation and benefits within an organization. The various market charges recovered from livestock dealers were registration fee, sales tax, toll tax and certificate charges are to be made at concessional rates.

6. Support Services

Information about support system is a must for an enterprise. In short-term it is the information which helps in-sound decision making. The information could be on infrastructure facilities, incentives available, financial tie-ups, availability of raw materials, tax concessions, *etc.* Innovative institutional and operational mechanisms and business models need to be developed to enable small-scale producers, especially women, to seize market opportunities along agricultural value chains, while taking into account issues of gender-based power inequalities and access to choices and resources. As livestock products are intrinsically difficult to trade because they are perishable, so require freezing, canning, drying, or processing of some sort to allow adequate shelf life to send to distant markets. Provision of State of Art- Abattoir cum meat processing plants and cold storage facilities at field level, establishment of custom hiring centres for the women, availability of quality inputs in bulk through authenticated sources can promote more women entrepreneur in this sector. Regional feed-fodder banks needs to be encouraged for milk production in summer months when most Indian cattle and buffaloes go dry.

7. Strengthening Women's Capacity and Leadership Skills

One of the most significant barriers faced by Indian women is their low level of literacy as education is key to achieving self-reliance and fighting poverty. Building women's capacity, especially leadership skills, is crucial to build women's self-confidence. Various organization have been developing tools and methodologies for strengthening the capacity of women to ensure that gender goals can be achieved and sustained in the long run. Capacity development enables women to access information about various organizations and to become members, and to participate more actively in their activities as well as decision making processes. On-farm technical trainings and extension services including trainings on veterinary health care, clean milk production, improved feeding practices involving local feed resources, better breeding and animal husbandry integrating the helpful traditional practices. In addition they also need the skills to run their own business. These include accounting skills, managerial skills and, marketing skills. Marketing management skills should be improved among the rural entrepreneurs to face the problems of entrepreneurship. Entrepreneurship development cell should be established at all the villages level to provide guidance and counselling to motivate the rural entrepreneurs regarding the use of modern technology. Training institutions supported by the Ministry of Agriculture and Farmers' Welfare, Ministry or Rural Development, Government of India (www. rural.nic.in) also impart skill and capacity building training which can be made use by these women entrepreurs.

8. Creating an Enabling Policy Environment

Governments play an important role in creating the enabling environment so that effective, inclusive and gender equitable organizations can flourish and act as important contributors to poverty reduction and the achievement of food security. An enabling environment will be a gender sensitive one when policies and legislation that directly or indirectly recognize that women and men from various social groups have differentiated roles and priorities; and therefore, adopt the appropriate measures so that all rural women and men can participate in, and benefit from emerging opportunities (FAO, 2000). Therefore various schemes and plans of government for the encouragement of women entrepreneurs should be strongly executed at different levels. Policies should be flexible to facilitate local circumstances. The nature of enterprises to be established in rural areas must be conducive to those areas in economic, social and environmental terms. Rural enterprise policy should cover all types of rural enterprise. There should be consistency and co-ordination with respect to the choice of rural enterprise locations.

Strategies and Approaches to Develop Women Entrepreneurs in Livestock Sector

Right efforts on from all areas are required in the development of women entrepreneurs and their greater participation in the livestock based entrepreneurial activities. Following strategies can be taken into account for effective development of women entrepreneurs in livestock sector.

- ✰ Consider rural women as specific target group for livestock based developmental programmers.
- ✰ Encourage women's participation in decision-making.
- ✰ Adequate trainings on management skills, professional competence and leadership skills to be extended to women community that enables them to understand the production process and production management.
- ✰ Activities in which women are trained should focus on their marketability and profitability.
- ✰ Counselling through the aid of committed NGOs, psychologists, managerial experts and technical personnel should be provided to existing and emerging livestock women entrepreneurs.
- ✰ Helping women farmers to identify their production, processing and market related constraints through awareness, exposure, exchange of information through other women farmers, extension officers and other stakeholders.
- ✰ Continuous contact with women farmers shall be established to provide knowledge and information empower and facilitate them to demand and access services; redress their concerns and grievances; and capture ground reality for use in policy making and improving delivery through; Farmers portal, Kisan Call Centers, Common Service Centers, Short message services and Community radio stations *etc.*
- ✰ Assisting women farmers to make best use of livestock technologies and good practices of improving livestock produce through capacity building.
- ✰ Establishing linkages through information source on livestock produce, new technologies and market related information such as, demand, supply and prices.
- ✰ Knowledge generators (public and private, formal and informal systems) should develop knowledge portals for capturing, collating and disseminating knowledge through all channels including print and electronic media, farm journals, regular meets *etc.* for the women farmers.
- ✰ Reaching out to and connecting women farmers and other stakeholders so that they communicate with each other.

- ☆ Test all indigenous technical know-how (ITKs), identify women farmer innovators, recognize their skills and leverage for field extension as farmer consultants.
- ☆ Promotion of Livestock based Farmers Field Schools for the women farmers
- ☆ Demand-Driven or Farmer-led and Market–Led Extension Approach and Farmer to Farmer Extension
- ☆ Introduction of ICT in delivery of extension messages and market information to the women farmers/entrepreneurs.
- ☆ Establishment of proper Research-Extension-Farmer-Market-Linkage for the existing and emerging livestock women entrepreneurs.
- ☆ A mechanism may be put in place to provide weekly/daily personalized information (thru SMS/MMS) to the women farmers through Veterinary Universities/Departments of Animal Husbandry/KVKs.
- ☆ A stakeholder directory of Institutions/Agencies, Experts/Livestock Magazines/Journals/Newspapers/channels shall be prepared and linked on to Farmers Portal.
- ☆ A comprehensive livestock-knowledge portal may be designed and developed to host information about the happenings/innovations in Livestock sector.
- ☆ The e-procurement of inputs and services required by the women farmers may be promoted to give them negotiating power to get the livestock inputs at the most competitive rates without transaction costs.
- ☆ Path-breaking methods such as Kala Jathas, Extension Buses can be tried successfully to deliver the message effectively on a large scale.
- ☆ Promotion of group approach
- ☆ Focus on nutritional value and new or unknown products
- ☆ The technical skills and ability of the women entrepreneurs should be evaluated
- ☆ Extension Agencies and Women Farmers Organizations should give wider publicity
- ☆ Encourage the women entrepreneurs by introducing various concessions and incentives.
- ☆ Networks of women entrepreneurs may be established
- ☆ Promotion of direct marketing by establishing close interaction between producers and consumers
- ☆ Promote organized road side display models milk and meat products for the existing and emerging livestock women entrepreneurs.

Conclusion

Gender mainstreaming and gender sensitization is a challenge but through livestock based enterprises/ventures in rural area can help to meet these challenges. These enterprises not only enhance national productivity, generate employment but also help to develop economic independence, personal and social capabilities like; economic empowerment, improved standard of living, improvement in leadership qualities, involvement in solving problems related to women, decision making capacity in family and community, build-up self confidence, increased social relations, political participation, increased participation in various developmental programs and enhance awareness *etc.* amongst rural women.

Livestock women entrepreneurs may contribute significantly in rural and economic development in India. However, lacks of suitable gender mainstreaming and gender sensitization policies, supportive network, financial and marketing prospects may hamper their entrepreneurial activities. National policies should be resolute in tackling this issue and local bodies should ensure the implementation of these policies at the community level and at last but not the least the rural women needs to be motivated to take up entrepreneurship as a career, with training and sustaining support systems providing all necessary assistance.

REFERENCES

Anon. 1997 Gender Development and NABARD: a brochure of NABARD. National Bank for Agricultural and Rural Development, Mumbai, India

Borkar, S.D., Joy, S., Lahamge, M. and Thakare, A. 2017. Women Empowerment in Veterinary Field: A Review. *International Journal of Scientific Research and Management*, 5(1): 5075-78.

Datta, T.N., and Ganguly, B.K. 2002. Analysis of consumer expenditure patterns in states with special reference to milk and milk products. National Information Network, National Dairy Development Board.

Devendra, C. and Chantalakhana, C. 2002. Animals, poor people and food security: opportunites for improving livelihoods through efficient natural resource management. *Outlook on Agriculture.* 31(3): 161-175.

FAO. 2000. Agriculture Tow (Rangnekar, 1994, Pandey, 2001), ANTHRA, 2005.

Garcia, O., Saha, A., Mahmood, K., Ndambi, A., and Hemme, T. 2006. Dairy development programs in Andhra Pradesh, India: impacts and risk for small-scale dairy farms. Pro-poor Livestock Policy Initiative Working Paper No. 38.

IFAD. 2009. Gender in agriculture sourcebook: Gender and livestock module – Overview. International Fund for Agricultural Development (IFAD), World Bank and the Food and Agriculture Organization of the United Nations, Rome.

Choudhary, K.N. 2011. Opportunities and challenges for rural women entrepreneurship in India. *Variorum Multi- Disciplinary e-Research Journal*, 1(3): 1-4.

Planning Commission Govt. of India. 2012. The Working Group on Animal Husbandry and Dairying - 12th Five Year Plan.

Qureshi, M.A., Khan, P.A. and Uprit, S. 2016. Empowerment of rural women through agriculture and dairy sectors in India. *Economic Affairs*, 61(1): 75-79.

Kumari, S., Kaushik, V. and Lodha, N. 2010. Problems Faced by Rural Women Entrepreneurs of Rajasthan. Stud Home Comm Sci, 4(2): 115-119

Waters-Bayer, A. 1985. Dairying by settled Fulani women in Central Nigeria and some implications for dairy development. Pastoral Development Network, 20. ODI, London.

Younas, M., Gulrez, S. and Rehman, H. 2007. Women's role in livestock production. The Dawn. Dec 17, 2007. http://www.dawn.com.

Index

D

E

F

G